Optimization of the Communication System for Networked Control Systems

Von der Fakultät Konstruktions-, Produktions- und Fahrzeugtechnik
der Universität Stuttgart zur Erlangung der Würde eines
Doktor-Ingenieurs (Dr.-Ing.) genehmigte Abhandlung

Vorgelegt von

Rainer Blind

aus Stuttgart

Hauptberichter: Prof. Dr.-Ing. Frank Allgöwer
Mitberichter: Prof. Karl Henrik Johansson
Prof. Dr. rer. nat. Kurt Rothermel

Tag der mündlichen Prüfung: 18. 12. 2013

Institut für Systemtheorie und Regelungstechnik
Universität Stuttgart
2014

Bibliografische Information der Deutschen Nationalbibliothek

Die Deutsche Nationalbibliothek verzeichnet diese Publikation in der
Deutschen Nationalbibliografie; detaillierte bibliografische Daten sind
im Internet über http://dnb.d-nb.de abrufbar.

ISBN 978-3-8325-3746-3

Logos Verlag Berlin GmbH
Comeniushof, Gubener Str. 47,
10243 Berlin
Tel.: +49 (0)30 42 85 10 90
Fax: +49 (0)30 42 85 10 92
INTERNET: http://www.logos-verlag.de

For my family.

Acknowledgments

This thesis contains the results obtained during my time as a research assistant at the Institute for Systems Theory and Automatic Control (IST), University of Stuttgart. I want to express my deepest gratitude to those who supported me and made this time highly enjoyable and unique.

First and foremost, I want to thank my PhD supervisor Prof. Dr.-Ing. Frank Allgöwer. His enthusiasm for control theory infected me and forced me to start my research activities at the IST. During this time, he gave me all the freedom I needed for my research and always supported me and believed in me. Moreover, he supported my trips to many international conferences throughout the world. He also gave me the opportunity to teach and thereby not only helped me to improve my didactic skills but also to deepen my understanding of control theory.

Second, I want to thank Prof. Karl Henrik Johansson and Prof. Dr. rer. nat. Kurt Rothermel for their interest in my research and being part of my doctoral exam committee.

Moreover, I want to thank Frank Allgöwer and all the members and guests of the IST for creating a very friendly and highly encouraging atmosphere. I enjoyed many interesting and inspiring presentations and discussions during my time at the institute. Most of them influenced only my thinking and research, but some discussions with Andreas Benzing, Mathias Bürger, Ben Carabelli, Georg Seyboth, Markus Kögel, Ulrich Münz, and Stefan Uhlich led to joint publications.

However, my time at the IST was not only limited to research. When looking back, I also remember with great joy and pleasure all the hikes during our Söllerhaus stays in the Kleinwalsertal as well as the rides to it and back home. Some of them where indeed epic. I also highly enjoyed the trips to the conferences, not only due to the inspiring discussions with colleagues, but also due to the fun we had during the coffee breaks, the welcome and farewell receptions, our leisure time activities, and the trips during the additional days before and afterwards. I want to thank all those who joined but also those who became friends despite different research interests and not a single joint hike, ride, or conference trip.

I would also like to thank the SPP 1305 *Control Theory of Digitally Networked Dynamical Systems* of the German Research Foundation (DFG) for the financial support and its members for the organization of workshops as well as the many interesting and fruitful discussions.

Special thanks go to Mathias Bürger, Jan Maximilian Montenbruck, Georg Seyboth, and Jingbo Wu for proofreading this thesis and our secretary team for helping me to overcome various bureaucratic hurdles. Moreover, I want to thank Heiko Seng for supporting me at most of my Ironman races as well as Johannes Reichart and Fabian Friedrichs for several swim sessions and many technical discussions.

Last but by no means least, I would like to thank my parents Gerd and Roswitha, my brother Thilo, my wife Nadja and our children Sophie, Jonas, and Eila Maria for all their support and love.

Stuttgart, July 2014
Rainer Blind

Table of Contents

Acronyms

ACK	Acknowledgement
CAN	Controller Area Network
CDF	Cumulative Distribution Function
CSMA	Carrier Sense Multiple Access
EB	Event-Based
FDMA	Frequency Division Multiple Access
iid	independent and identically distributed
ISO	International Organization for Standardization
LLC	Logical Link Control
MAC	Medium Access Control
MARE	Modified Algebraic Riccati Equation
MDC	Multiple Description Coding
NAK	negative acknowledgement
NCS	Networked Control System
OSI	Open Systems Interconnection
PDF	Probability Density Function
PMF	Probability Mass Function
QoS	Quality of Service
TCP	Transmission Control Protocol
TDMA	Time Division Multiple Access
TT	Time-Triggered
UDP	User Datagram Protocol
WLAN	Wireless Local Area Network

Abstract

Networked Control Systems are control systems, where the feedback loop is closed by a communication system. Within the past decades, the effects of the properties of the communication system, like loss, delay, or bandwidth constraints, on the control performance have been studied thoroughly. However, in the field of communication theory, it is well known that the properties of the communication system depend on the design of the communication system *and* its usage. When taking this into account, the proper design of a networked control system becomes a very challenging task because the controller must not only cope with the loss and delay of the communication system but might also be responsible for it.

The goal of this thesis is to work towards a joint design of the controller and the communication system for networked control systems. To achieve this goal, we combine methods from control and communication theory. We build upon previous works from the field of networked control systems, where controller design methods for a communication system with given properties are presented, but take into account that these properties depend on the design of the communication system and its usage. Using some well known ideas from communication theory, we derive several methods to improve the control performance by optimizing the communication system.

First, we present two approaches to improve remote estimation over a communication system with packet loss. For a system with two or more measurements, we show that this can be achieved by a linear precoding, i.e., a linear transformation, of the measurements before sending them over the communication system. Moreover, we show that by retransmitting lost measurements, it is possible to achieve optimal estimates, although retransmitted measurements will be outdated on their arrival. Next, optimal control over a communication system with packet loss and delay is studied. By increasing the sampling time to allow several transmissions during each sampling interval, it is possible to increase the reliability of the communication system. However, increasing the sampling time generally degrades the performance, such that the optimal choice of the sampling time becomes an interesting problem, which is considered in this thesis. When the communication system is actually a network, loss and delay depend on the route through this network. Since the control performance depends on the loss and delay of the communication system, we consider the optimal routing through the communication system. The details of the interaction between control and communication are finally studied by analyzing and comparing time-triggered and event-based control over a shared communication system.

Abstract
_____

Deutsche Kurzfassung

Motivation

Nachdem das Internet Ende der 1990er Jahre seine Praxistauglichkeit bewies und allgemeine Bekanntheit und Beliebtheit erhielt, wird immer häufiger daran gearbeitet, Regelkreise mit Hilfe eines paketbasierten Kommunikationssystems zu schließen. Dieser Trend führte zu dem neuen Forschungszweig der *digital vernetzten Regelung* (engl. *Networked Control System*). Durch die Nutzung eines paketbasierten Kommunikationssystems können die vorhandenen Ressourcen effizient geteilt werden, wodurch der Verkabelungsaufwand reduziert und eine billige, flexible und schnelle Datenübertragung ermöglicht wird. Die Verwendung eines paketbasierten Kommunikationssystems hat allerdings auch einen Nachteil: Datenpakete können verloren gehen oder stark verzögert werden. Regelungstechniker müssen also lernen mit Paketverlust und -verzögerung umzugehen. In den letzten Jahren wurden deshalb zahlreiche Methoden vorgestellt, mit denen die Stabilität des geschlossenen Kreises sowie eine hohe Regelgüte auch bei Paketverlust und/oder -verzögerung garantiert werden können. Bei diesen Methoden handelt es sich überwiegend um Methoden zum Reglerentwurf. Die Möglichkeit, das Kommunikationssystem so zu entwerfen, dass es für die speziellen Anforderungen der Regelung besonders geeignet ist, wurde bisher selten verfolgt. Da dieser Ansatz jedoch eine interessante und vielversprechende Möglichkeit ist, die Regelgüte von digital vernetzten Systemen weiter zu steigern, werden in dieser Arbeit verschiedene Methoden zur Optimierung des Kommunikationssystems bei digital vernetzten Systemen vorgestellt.

Ursprünglich wurde das Internet hauptsächlich für den Austausch von Dateien optimiert, was zwar eine zuverlässige Datenübertragung erfordert, die Übertragungsdauer aber relativ unwichtig ist. In letzter Zeit werden aber immer häufiger auch Echtzeitdaten für Sprache, Musik und sogar Filme über das Internet übertragen. Im Vergleich zum Austausch von Dateien sind bei solchen Übertragungen die Anforderungen an das Echtzeitverhalten deutlich höher, aber die Anforderungen bezüglich Zuverlässigkeit geringer. Wird ein Regelkreis von einem Kommunikationssystem geschlossen, dann sind sowohl eine hohe Zuverlässigkeit als auch eine geringe Übertragungszeit wichtig. Weil ein ideales Kommunikationssystem ohne Paketverlust und -verzögerung sehr schwer, oder gar unmöglich, zu realisieren ist, werden im Bereich der digital vernetzten Regelung Methoden für die Regelung über ein Kommunikationssystem mit Paketverlust und -verzögerung entwickelt. Dabei wird der Regler meist für ein Kommunikationssystem mit gegebenen Anforderungen bezüglich Verlust, Verzögerung und verfügbarer

3

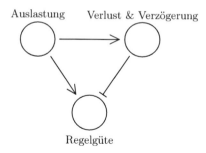

Abbildung 1.: Auswirkung von Auslastung, Paketverlust und -verzögerung auf die Regelgüte.

Bandbreite entworfen, ohne dabei zu berücksichtigen, dass diese Eigenschaften vom Kommunikationssystem und dessen Nutzung abhängen könnten. Im Bereich der Kommunikationstechnik ist jedoch bekannt, dass Paketverlust und -verzögerung nicht unveränderlich und unabhängig voneinender sind sondern vom Kommunikationssystem und dessen Nutzung abhängen, siehe z. B. Rom and Sidi (1990); Tanenbaum (2003). Weil der Regler für die Nutzug des Kommunikationssystems verantwortlich ist, dieser aber auch mit den Folgen einer nicht perfekten Kommunikation umgehen muss, ergibt sich also eine komplexe Interaktion von Regelung und Kommunikation. Abbildung 1 veranschaulicht dies und zeigt, wie sich die gesendete Datenmenge sowie Paketverlust und -verzögerung auf die Regelgüte auswirken. Im Allgemeinen wird die Regelgüte durch eine Erhöhung der gesendeten Datenmenge verbessert, d. h. je mehr Daten übertragen werden, desto besser wird die Regelgüte. Andererseits verschlechtert sich die Regelgüte bei Paketverlust und -verzögerung, d. h. je mehr Paketverlust und -verzögerung, desto schlechter wird die Regelgüte. Darüber hinaus ist im Bereich der Kommunikationstechnik bekannt, dass Paketverlust und -verzögerung mit der Auslastung des Kommunikationsnetzes ansteigen, aber auch von anderen Faktoren, wie z. B. dem Kommunikationsprotokoll und dem Verkehrsmuster, abhängen. Außerdem existieren im Bereich der Kommunikationstechnik verschiedene Methoden um Paketverlust durch Paketverzögerung zu kompensieren, z. B. durch das erneute Senden von verloren Paketen oder durch die Wahl einer anderen Route durch das Kommunikationssystem. Diese Zusammenhänge sind relativ einfach und verständlich solange sie einzeln betrachtet werden. Wenn jedoch diese Zusammenhänge gleichzeitig berücksichtigt werden, dann wird das Zusammenspiel von Regelung und Kommunikation sehr komplex und der gemeinsame Entwurf von Regler und Kommunikationssystem zur Herausforderung.

Für die Realisierung einer digital vernetzten Regelung muss aber nicht nur der Regler entworfen werden, sondern auch das Kommunikationssystem. Zur Zeit werden diese meist unabhängig voneinenader entworfen, was nicht immer der beste Ansatz ist.

Wie bereits erwähnt, werden im Bereich der digital vernetzten Regelung überwiegend Reglerentwurfsmethoden für ein Kommunikationssystem mit gegebenen Eigenschaften publiziert. Dadurch wurden die Auswirkungen der Eigenschaften des Kommunikationssystems (Verlust, Verzögerung, Beschränkung der Bandbreite, ...) auf die Regelgüte schon sehr ausführlich untersucht, selten dagegen die Auswirkungen der Eigenschaften des Reglers (Abtastzeit, Abtaststrategie, ...) auf das Kommunikationssystem. Noch seltener wurde das Zusammenspiel von Regelung und Kommunikation untersucht. Für weitere Fortschritte im Bereich der digital vernetzten Regelung ist es entscheidend, diese Kluft zwischen Regelungs- und Kommunikationstechnik zu schließen und den Regler zusammen mit dem Kommunikationssystem zu entwerfen. D. h. die Separation zwischen dem Entwurf des Reglers und dem Entwurf des Kommunikationssystems muss überwunden werden. Beim Entwurf eines digital vernetzten Reglers wird also auch detailliertes Wissen über das verwendete Kommunikationssystem benötigt. Dieses Wissen ist aber nicht nur nötig um Fehler zu vermeiden, sondern kann auch dazu genutzt werden, die Regelgüte zu verbessern. In dieser Arbeit soll deshalb gezeigt werden, wie verschiedene Methoden aus dem Bereich der Kommunikationstechnik verwendet werden können, um die Regelgüte zu verbessern.

Kommunikationssysteme

Um die immer größer werdende Komplexität der Kommunikationssysteme in einer strukturierten Art und Weise zu handhaben, wurde in Zimmermann (1980) ein geschichteter Aufbau des Kommunikationssystems vorgestellt, welcher heute als Schichtenmodell bzw. ISO/OSI Referenzmodell bekannt ist. Die grundlegende Idee dieses Schichtenmodells ist folgende: Jede Schicht ist für einige wohldefinierte Dienste zuständig. Zur Realisierung dieser Dienste werden die Dienste der darunterliegenden Schicht genutzt. D. h. Schicht N nutzt die Dienste von Schicht $N-1$ um Schicht $N+1$ komplexere Dienste anzubieten. Dieses Schichtenmodell erlaubt es, eine zuverlässige Ende-zu-Ende Verbindung über das Internet zu bilden, ein im Grunde unzuverlässiges, weltweites Kommunikationssystem mit einer Vielzahl von verschiedenen Technologien. Obwohl die Schichten und Dienste von heutigen Kommunikationssystemen teilweise stark von dem ursprünglichen Modell aus Zimmermann (1980) abweichen, stellt die zugrunde liegende Idee immer noch den Grundpfeiler heutiger Kommunikationssysteme dar.

Abbildung 2 zeigt die sieben Schichten des ISO/OSI Referenzmodells. Der Nutzer eines Kommunikationssystems interagiert mit der obersten Schicht, der *Anwendungsschicht* (engl. *Application Layer*). Die *Darstellungsschicht* (engl. *Presentation Layer*) ermöglicht eine systemunabhängige Darstellung der Daten. Zu den Diensten der Darstellungsschicht gehört deshalb die Konvertierung von systemabhängigen Daten, aber auch die Kompression und Verschlüsselung von Daten. Im ISO/OSI Modell stellt die *Sitzungsschicht* (engl. *Session Layer*) Dienste zum Auf- und Abbau von Verbindungen sowie deren Management zur Verfügung. Eine Verbindung zwischen zwei Endknoten,

Anwendung
Darstellung
Sitzung
Transport
Vermittlung
LLC · · · · · · Sicherung MAC
Bitübertragung

Abbildung 2.: Die sieben Schichten des ISO/OSI Referenzmodells.

d.h. eine Ende-zu-Ende Verbindung, wird durch die *Transportschicht* (engl. *Transport Layer*) realisiert. Zur Realisierung einer zuverlässigen Ende-zu-Ende Verbindung kann die Transportschicht Dienste wie eine Ende-zu-Ende Flusskontrolle, Überlastkontrolle und Fehlerkorrektur zur Verfügung stellen. Die Flusskontrolle verhindert, dass der Sender schneller Daten sendet, als der Empfänger verarbeiten kann. Eine Überlastkontrolle verhindert eine Überlastung der Verbindungen des zugrunde liegenden Kommunikationssystems. Die Fehlerkorrektur garantiert, dass alle gesendeten Daten korrekt empfangen werden. Innerhalb der Transportschicht wird dies meist durch den Versand von Bestätigungen (engl. acknowledgement) für korrekt empfangene Pakete und einer erneuten Übertragung von unbestätigten Paketen realisiert. Innerhalb der Internetprotokollfamilie sind das Transmission Control Protocol (TCP) und das User Datagram Protocol (UDP) die bekanntesten Protokolle der Transportschicht. Die *Vermittlungsschicht* (engl. *Network Layer*) ist hauptsächlich für die Wahl der Route zwischen Start- und Endknoten zuständig. Da komplexe Netzwerke wie das Internet meist aus vielen verschiedenen kleineren Netzwerken zusammengesetzt sind, gehört zu den Diensten der Vermittlungsschicht aber auch die Überwindung dieser Differenzen. Die *Sicherungsschicht* (engl. *Data Link Layer*) ist verantwortlich für die Übertragung von Daten zwischen zwei Knoten eines Netzwerkes sowie die Erkennung und Behebung von Fehlern, die in der Bitübertragungsschicht auftreten. Um dies zu ermöglichen, wird die Sicherungsschicht in zwei Unter-Schichten unterteilt: *Logical Link Control (LLC)* und *Medium Access Control (MAC)*. Die LLC Schicht ist für die Flusskontrolle und Fehlerkorrektur zwischen zwei Knoten verantwortlich. Die MAC Schicht definiert, wie auf das physikalische Medium zugegriffen wird, d.h. diese Schicht regelt, wann ein Nutzer senden darf. Die bekanntesten Protokolle der MAC Schicht sind ALOHA und verschiedene Variationen von CSMA (engl. Carrier Sense Multiple Access) wie z.B. Ethernet oder WLAN. Die *Bitübertragunsschicht* (engl. *Physical Layer*) definiert wie die Bits auf dem physikalischen Medium dargestellt werden, z.B. durch eine bestimmte Spannung oder einen Lichtimpuls.

Forschungsbeiträge und Gliederung der Arbeit

Um die Regelgüte von digital vernetzten Systemen weiter zu steigern, ist es notwendig, das Verständnis über das Zusammenspiel von Regelung und Kommunikation zu vertiefen. Dazu muss die Kluft zwischen Regelungs- und Kommunikationstechnik geschlossen werden, d.h. die Trennung zwischen Regelung und Kommunikation muss aufgehoben werden. Um dieses Ziel zu erreichen, muss beim Reglerentwurf die Annahme von gegebenen Eigenschaften des Kommunikationssystems verworfen werden. Deshalb werden in dieser Arbeit verschiedene Methoden zum gemeinsamen Entwurf von Regler und Kommunikationssystem vorgestellt. Insbesondere wird gezeigt, wie die Regelgüte durch eine Optimierung des Kommunikationssystems verbessert werden kann. Diese Methoden sind in folgenden Schichten des ISO/OSI Modells eingeordnet.

Darstellungsschicht In Kapitel 2 wird gezeigt, dass durch eine lineare Vorkodierung (engl. linear precoding) die Zustandsschätzung über ein verlustbehaftetes Kommunikationssystem verbessert werden kann. Dazu wird ein System mit zwei Ausgängen betrachtet, bei dem die Messdaten über zwei voneinander unabhängige, verlustbehaftete Verbindungen zu einem Kalman Filter gesendet werden. Bevor die Messdaten über das Kommunikationssystem gesendet werden, werden sie durch eine lineare Vorkodierung transformiert. Dadurch ist es möglich, die Zustandsschätzung oder die Robustheit des Kalman Filters gegen Paketverlust zu verbessern. Dieses Kapitel basiert auf Blind et al. (2009).

Transportschicht In Kapitel 3 und 4 wird der Frage nachgegangen, ob es sich lohnt, verlorene Pakete erneut zu senden. Dazu wird zunächst in Kapitel 3 das erneute Senden von unbestätigten Messdaten betrachtet. Weil wiederholt gesendete Messdaten bei ihrer Ankunft jedoch bereits veraltet sind, ist es natürlich fraglich ob dieser Ansatz sinnvoll ist. In Kapitel 3 wird deshalb gezeigt, dass durch das erneute Senden von verlorenen Messdaten eine optimale Zustandsschätzung erreicht werden kann, ohne die Nachteile bisheriger Methoden in Kauf nehmen zu müssen, bei denen eine optimale Zustandsschätzung nur durch die Berechnung der Zustandsschätzung am Sensor oder dem Versand von sehr großen Datenpaketen, die alle Messdaten enthalten, erreicht wird. Dieses Kapitel basiert auf Blind and Allgöwer (2013a).

Wie bereits erwähnt, sind in Kapitel 3 die erneut gesendeten Messdaten veraltet, wenn sie ankommen. In Kapitel 4 wird deshalb vorgeschlagen, die Abtastzeit lang genug zu wählen damit innerhalb eines Abtastintervalls ein Datenpaket mehrfach gesendet werden kann, oder es im Falle eines Verlustes erneut gesendet werden kann. Dies führt zu einem interessanten Problem: Das erneute Senden von verloren Paketen reduziert zwar die Verlustwahrscheinlichkeit, erfordert aber die Abtastzeit zu erhöhen. Weil die Regelgüte im Allgemeinen durch eine Reduktion der Verlustwahrscheinlichkeit verbessert, aber durch eine Verlängerung der Abtastzeit verschlechtert wird, ist es nicht klar, ob es sich lohnt, die Abtastzeit zu verlängern, um das erneute Senden von Paketen zu ermöglichen. Deshalb wird in Kapitel 4 die optimale Regelung eines zeitkontinuierlichen Systems über ein Kommunikationssystem mit Paketverlust und -verzögerung betrachtet und untersucht, wie sich die Abtastzeit auf die Regelgüte auswirkt. Anhand von drei Beispielen wird gezeigt, dass durch das Design der Transportschicht und eine geeignete Wahl der Abtastzeit sowohl der Bereich, für den ein stabilisierender Regler existiert, vergrößert werden kann, als auch die Regelgüte verbessert werden kann. Dieses Kapitel basiert auf Blind and Allgöwer (2013d); eine frühere Version wurde in Blind and Allgöwer (2012b) veröffentlicht.

Vermittlungsschicht In Kapitel 5 wird angenommen, dass die Route durch das Kommunikationssystem gewählt werden kann, wodurch sich die Möglichkeit ergibt, diese so zu wählen, dass die Regelgüte optimiert wird. Um diese Aufgabenstellung mathematisch exakt zu formulieren, wird das zugrunde liegende Kommunikationssystem als Graph modelliert. Jede Kante des Graphen repräsentiert eine Verbindung des zugrunde liegenden Kommunikationssystems. Jede Verbindung verliert Pakete mit einer gewissen Wahrscheinlichkeit und benötigt für deren Übertragung eine gewisse Zeit. Damit ist die Ende-zu-Ende Verzögerung die Summe der Verzögerungen der verwendeten Verbindungen. Die Ende-zu-Ende Ankunftswahrscheinlichkeit ergibt sich ebenfalls aus den Ankunftswahrscheinlichkeiten der verwendeten Verbindungen. Durch diese mathematische Beschreibung des Kommunikationssystems ist es möglich, das Problem der optimalen Regelung und Routenwahl als ein Optimierungsproblem mit ganzzahligen Zwangsbedingungen zu schreiben. Dieses Kapitel basiert auf Blind and Allgöwer (2013c).

MAC Schicht Das Zusammenspiel von Regelung und Kommunikation innerhalb der MAC Schicht wird in Kapitel 6 untersucht. Dazu wird zunächst die ereignisbasierte und zeitgetriggerte Regelung mit einem gemeinsam genutzten Kommunikationssystem analysiert und dann diese beiden Ansätze miteinander verglichen. Beim Vergleich dieser beiden Abtaststrategien fällt insbesondere der Unterschied im generierten Verkehr auf. Bei der zeitgetriggerten Regelung haben die Abtastzeiten einen konstanten Abstand und sind im Voraus bekannt. Bei der ereignisbasierten Regelung ergeben sich dagegen keine konstanten Abstände zwischen den Ereignissen und es ist auch nicht im Voraus bekannt, wann ein Ereignis eintritt. Dieser Unterschied im Verkehrsmuster beeinflusst die Wahl des Kommunikationssystems sowie dessen Verlust und Verzögerung. Für eine genaue Analyse von ereignisbasierter und zeitgetriggerter Regelung mit einem gemeinsam genutzten Kommunikationssystem ist es also notwendig, die Details des Medienzugriffs zu berücksichtigen. Deshalb wird die ereignisbasierte Regelung mit klassischen wettbewerbsbasierten Zugriffsverfahren, ALOHA und verschiedene Variationen von CSMA, sowie die zeitgetriggerte Regelung mit den zwei bekanntesten deterministischen Zugriffsverfahren, TDMA und FDMA betrachtet. Dieses Kapitel basiert auf Blind and Allgöwer (2013b); frühere Versionen sind in Blind and Allgöwer (2011a,b,c) veröffentlicht.

Chapter 1.

Introduction

1.1. Motivation

Since the raise of the Internet in the late 1990s, there is a steadily increasing interest to use a packet based communication system to close the feedback loop, giving birth to the new field of *Networked Control Systems (NCS)*. Using a packet based communication system allows to efficiently share resources and thereby reduces the wiring harness and provides a cheap, flexible, and fast data transmission. As a drawback, a packet based communication system might loose packets or delay them significantly. Consequently, control engineers are forced to deal with packet loss and delay and have to guarantee stability of the closed loop as well as a good control performance despite these effects. Thus, many controller design methods for networked control systems have been published in the last decade. Much less effort has been invested in the design of the communication systems for networked control systems.

Originally, the Internet was designed mainly for file transfers, which must be reliable but the transmission time is not critical. Recently, also real time data for, e.g., voice, music, and even video, is transmitted over the Internet. The transmission times for this real time data is much more critical but the connection does not need to be as reliable as when transmitting a file. When a control loop is closed by a communication system, a reliable and timely connection is crucial. Since an ideal connection without loss and delay would be very difficult, or even impossible, to realize, the control over a communication system with loss and delay is studied in the field of networked control systems. Thereby, the controller is most often designed for a communication system with given properties, like loss probability, delay characteristics, or bandwidth constraints, without taking into account that these properties might depend on the design of the communication system and its usage. When browsing through the literature on communication systems, e.g., Rom and Sidi (1990); Tanenbaum (2003), it becomes clear that loss and delay are not fixed and independent of each other but strongly depend on the communication system and its usage. As a consequence thereof, the joint design of control and communication is not straightforward due to the complex interaction between control and communication. Figure 1.1 depicts how load, loss, and delay affect the control performance. In general, the control performance increases with the network load, i.e., the more data exchanged, the better the control perfor-

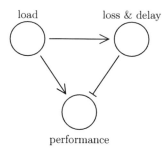

Figure 1.1.: The effect of load, loss, and delay on the control performance.

mance. On the other hand, the control performance decreases with loss and delay, i.e., the more loss and delay, the worse the control performance. Moreover, in the field of communication systems, it is well known that loss and delay increase with the network load but also depend on several other factors like the communication protocol and the traffic pattern. Moreover, there exist different methods to trade loss against delay, e.g., by retransmitting lost packets or choosing another route through the communication system. These relationships are relatively simple and well understood when considered separately. However, when taking all of them into account, the interaction between control and communication, and thus the joint design of control and communication, becomes a challenging problem.

To realize a networked control system, not only the controller, but also the communication system, must be designed. Currently, the design of the communication system and the design of the controller are most often studied separately, which is not necessarily the best approach. In the field of networked control systems, the effect of the properties of the communication system (loss, delay, bandwidth constraints, ...) on the control performance are already very well studied, while the effects of the properties of the controller (sampling time, sampling strategy, ...) on the communication system are seldom studied. The interaction between control and communication is studied even less. For further improvements in the field of networked control systems it is crucial to close this gap between control and communication theory and study the controller and communication system together. This essentially means that the separation between the design of the controller and the design of the communication system must be overcome. Hence, when designing a networked control system, the details of the communication system must be taken into account. However, a detailed knowledge about the communication system is not only necessary to avoid pitfalls. Instead, by adapting the methods used in communication theory, it is even possible to improve the control performance, as demonstrated within this thesis.

Application
Presentation
Session
Transport
Network
LLC ⋯⋯⋯ MAC Data Link
Physical

Figure 1.2.: The ISO/OSI Stack.

1.2. Communication System

To handle the ever increasing complexity of communication systems in a structured way, a layered architecture was presented in Zimmermann (1980), which is now well known as the ISO/OSI reference model. The basic idea of the ISO/OSI model is that each layer is responsible for a limited set of services and builds on the services of the layer below to implement these services, i.e., layer N uses the services of layer $N − 1$ to offer more powerful services to layer $N + 1$. This layered architecture allows the creation of a reliable end-to-end connection over an essentially unreliable, world wide communication system with a huge variety of technologies like the Internet. Although the layers and the services of todays communication systems differ from the ones defined in Zimmermann (1980), the basic idea of a layered architecture remains the cornerstone of todays communication systems.

Figure 1.2 depicts the seven layers of the ISO/OSI stack. The user of a communication system interacts with the topmost layer, the *Application Layer*. The *Presentation Layer* is responsible for the data representation and the conversion of machine dependent representations like character encodings and file formats but also encoding/decoding, encryption, and compression. Within the ISO/OSI reference model, the *Session Layer* is responsible for establishing, managing, and terminating connections. The *Transport Layer* enables end-to-end connections between two users. Thereby, it may provide services like end-to-end congestion control and end-to-end error control. Congestion control is the ability to avoid overloading the links of the underlying communication system. Error control in the transport layer is generally realized by acknowledging successfully transmitted packets and retransmitting non-acknowledged packets. In doing so, a reliable end-to-end connection can be realized although the underlying communication system might be unreliable. Within the Internet protocol suit, the two best known transport layer protocols are the Transmission Control Protocol (TCP) and the User Datagram Protocol (UDP). The *Network Layer* is responsible for the creation of a route between a source and a destination node. Since huge networks like the Internet are composed of many different networks, one challenge within the network layer is to overcome these differences. The *Data Link Layer* provides data transfer between two nodes of the same network as well as detection and correction of errors that occur in the physical layer. To fulfill these tasks, this layer is subdivided into the *Logical Link Control (LLC)* and *Medium Access Control (MAC)* Layer. The LLC Layer provides node-to-node flow and error control. The MAC Layer defines the medium access, i.e., this layer defines how and when a user is allowed to send. The best known MAC protocols are ALOHA and variations of Carrier Sense Multiple Access (CSMA) like Ethernet or WLAN. Finally, the *Physical Layer* defines how bits are represented on the medium, e.g., by a certain voltage or a light impulse.

1.3. Contribution of the Thesis

To further improve the performance of networked control systems, it is necessary to deepen our understanding of the interaction between control and communication and work towards closing the gap between control and communication. To achieve this goal, the separation between control and communication as well as the assumption of fixed properties of the communication system must be given up. Hence, we consider the joint design of the controller and the communication system and present several approaches to optimize the communication system to achieve a better control performance. These methods are located within the following layers of the ISO/OSI reference model:

Presentation Layer In Chapter 2, we use a technique called linear precoding to transform the measurements before sending them over the communication system. In doing so, it is possible to improve the state estimates of a Kalman filter with intermittent observations, or its robustness against packet loss, for a system with two or more outputs. This chapter is based on Blind et al. (2009).

Transport Layer In Chapter 3 and 4, we raise the question whether it is worth to retransmit lost packets. Thereby, we first consider the retransmission of measurement packets in Chapter 3. When retransmitting a measurement after a packet loss, this measurement will be outdated on its arrival. Thus, previous literature found no benefit in retransmitting lost packets. Nevertheless, we present an approach that achieves optimal state estimates and thereby show that retransmitting lost measurements is beneficial when done properly. This chapter is based on Blind and Allgöwer (2013a).

In Chapter 4, we suggest to choose the sampling time long enough to transmit a packet several times or retransmit lost ones within each sampling interval. This leads to an interesting tradeoff: retransmitting lost packets reduces the loss probability but requires to increase the sampling time. Since the control performance generally increases with a lower packet loss probability but decreases with a longer sampling time, it is not clear whether it is worth to increase the sampling time to allow packet retransmissions. This chapter is based on Blind and Allgöwer (2013d); an earlier version is published in Blind and Allgöwer (2012b).

Network Layer In Chapter 5, we assume that the route through the underlying communication system can be chosen in order to optimize the control performance. To solve this problem mathematically, we model the underlying communication system as a graph, where each edge of the graph represents a link of the underlying communication system. Each link of the underlying communication system transmits packets after some delay and with some probability. Now, the end-to-end delay of a route is just the sum of the delays of its links; the end-to-end arrival probability of a route also follows from the arrival probabilities of its links. By using this mathematical description of the underlying communication system, it is possible to write the optimal control and routing problem as an optimization problem with integer constraints. This chapter is based on Blind and Allgöwer (2013c).

MAC Layer The interaction between control and communication within the MAC layer is analyzed in Chapter 6. Therefore, we first analyze event-based and time-triggered control with a shared communication system and then compare these two approaches. When comparing time-triggered and event-based control, the difference of the traffic pattern is outstanding and affects the choice of the communication system and also its loss and delay. Thus, the details of the medium access must be taken into account when analyzing and comparing event-based and time-triggered control over a shared communication system. Hence, we study event-based control with the classic contention based protocols: ALOHA and several variations of CSMA. On the other hand, we study time-triggered control with the two most well known deterministic protocols Time Division Multiple Access (TDMA) and Frequency Division Multiple Access (FDMA). This chapter is based on Blind and Allgöwer (2013b); earlier versions are published in parts in Blind and Allgöwer (2011a,b,c).

Chapter 2.

Optimization within the Presentation Layer: Linear Precoding

The main service of the presentation layer is the translation of different representations of data. For file transmission, this means translating different character encodings or encryption and decryption. In this chapter, we build on this idea and present an approach to improve remote estimation over a lossy communication system by transforming the measurements before sending them over the communication system. This chapter is based on Blind et al. (2009).

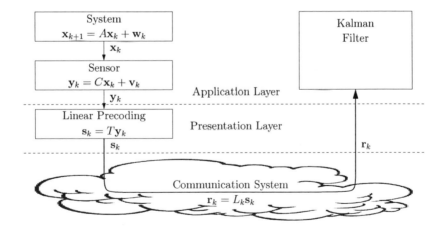

Figure 2.1.: Linear precoding to improve Kalman filtering over lossy links.

2.1. Motivation

In this chapter, we present a scheme to improve Kalman filtering over a communication system with random packet losses. Figure 2.1 depicts the considered setup, which will be described in detail in Section 2.2. For now, note that a linear precoding is performed before the measurements are sent over the communication system. This approach is best motivated by the following simple but illustrative example.

Suppose, the state of a discrete-time system

$$\mathbf{x}_{k+1} = \begin{bmatrix} 2.5 & 0 \\ 0 & 2 \end{bmatrix} \mathbf{x}_k,$$

shall be estimated over a communication system with random packet loss. Moreover, both states can be measured directly, i.e.,

$$\mathbf{y}_k = \begin{bmatrix} y_{k,1} & y_{k,2} \end{bmatrix}^\mathsf{T} = \mathbf{x}_k.$$

Each measurement is sent in an individual packet over the communication system, i.e., $y_{k,1}$ is sent in one packet and $y_{k,2}$ in another packet. Obviously, this system is observable as long as both measurements arrive but it is not observable, when one of the measurements is lost.

Now, suppose that the transformed measurements $s_{k,1} = y_{k,1} + y_{k,2}$ and $s_{k,2} = y_{k,1} - y_{k,2}$ are sent instead. Since this transformation is invertible, there is no difference to the original system when both measurements arrive. However, the system is still observable when only one of the transformed measurements arrives. Since observability is crucial for state estimation, we expect that a remote estimator with random packet loss will perform better when these transformed measurements are sent instead of the original measurements.

So far, this transform is motivated by system theoretic insight, namely the loss of observability. Interestingly, such a transform is very well known in the area of *Multiple Description Coding* (MDC) where it is called *linear precoding* or correlating transform. MDC aims at transmitting information over a communication system with random packet loss, called erasure channel, see Goyal (2001) for an overview of MDC. One possibility to cope with such an erasure channel is the usage of a correlating transform, see, e.g., Goyal and Kovacevic (2001); Romano et al. (2005); Uhlich and Yang (2008). In classical MDC, the correlating transform is chosen such that the mean squared error between the originally sent signal and the reconstructed signal is minimized. For the considered setup, this means that the measurements are reconstructed as well as possible. However, for state estimation we are more interested in a small error between the state and its estimate than a good reconstruction of the measurements. Thus, we adopt the idea of a correlating transform but design it to minimize the estimation error of the remote Kalman filter or maximize its robustness against measurement losses.

MDC to improve remote estimation is also considered in Jin et al. (2006). However, in Jin et al. (2006), scalar quantizers for MDC are used to code the measurements to minimize the distortion for a given bitrate. This increases the robustness of the system as the number of available bits are split into two independent streams that are transmitted. If only one of the two streams is available, then the original measurement can still be approximately recovered. Since classical MDC is used to encode the measurements, a good reconstruction of the measurements is achieved but not necessarily the best state estimates.

2.2. Problem Setup

2.2.1. Control System

We use the problem setup depicted in Figure 2.1 to demonstrate the benefit of pre-coding the measurements before sending them over a lossy communication system to a remote estimator. The goal is to estimate the state of the following discrete-time system over a communication system with random packet loss

$$\mathbf{x}_{k+1} = A\mathbf{x}_k + \mathbf{w}_k, \tag{2.1a}$$
$$\mathbf{y}_k = C\mathbf{x}_k + \mathbf{v}_k, \tag{2.1b}$$

where $\mathbf{x}_k \in \mathbb{R}^{n_x}$ is the system state and $\mathbf{y}_k \in \mathbb{R}^{n_y}$, $n_y \geq 2$ the measurement output at time instance k. Moreover, $\mathbf{w}_k \in \mathbb{R}^{n_x}$ and $\mathbf{v}_k \in \mathbb{R}^{n_y}$ are Gaussian white noise vectors with zero mean and covariance matrix $W \in \mathbb{R}^{n_x \times n_x}$, $W \geq 0$, and $V \in \mathbb{R}^{n_y \times n_y}$, $V > 0$, respectively. Moreover, $\mathbf{w}_k$ and $\mathbf{w}_s$ are independent for $s \neq k$, the pair (A, C) is detectable, and $(A, W^{\frac{1}{2}})$ is stabilizable.

2.2.2. Communication System

The model of the underlying communication system is relatively simple. At each sampling instance, the sensor passes a vector $\mathbf{s}_k \in \mathbb{R}^{n_s}$ to the communication system, which sends each component of the vector $\mathbf{s}_k$ in an individual packet. We assume that packet loss is iid (independent and identically distributed) and each packet arrives with probability p. Packet loss is modeled by a left-multiplication of the vector $\mathbf{s}_k$ with an erasure matrix $L_k \in \{0, 1\}^{n_{r,k} \times n_s}$, $n_{r,k} \leq n_s$, which is the identity matrix where a row is removed if the corresponding packet is lost. The following two examples will give the basic idea of this notation. Obviously, L_k is the identity matrix when all components arrive. If there are three components and at time k the first and third arrive, then we have $n_s = 3$, $n_{r,k} = 2$, and $L_k = \begin{bmatrix} 1 & 0 & 0 \\ 0 & 0 & 1 \end{bmatrix}$. Using this notation, the receiver receives the vector

$$\mathbf{r}_k = L_k \mathbf{s}_k, \quad \mathbf{r}_k \in \mathbb{R}^{n_{r,k}}. \tag{2.2}$$

Since the receiver knows which packets arrived at time instance k, the erasure matrix L_k is known by the receiver.

We define the set of all possible erasure matrices as $\mathcal{L} := \{\tilde{L}_1, \ldots, \tilde{L}_E\}$, where $E = |\mathcal{L}| = 2^{n_s}$ is the total number of erasure matrices. Since packet loss is assumed to be iid with arrival probability p, the probability that a particular erasure matrix $\tilde{L}_e$ is chosen from $\mathcal{L}$ is $w_e(p) := p^i(1-p)^{n_s-i}$, where i is the number of arrived components.

2.3. Linear Precoding

Instead of sending the measurements $\mathbf{y}_k$ directly over the communication system, they are first transformed by a linear precoding. This linear precoding is done by a left-multiplication with the precoding matrix $T \in \mathbb{R}^{n_s \times n_y}$. Thus, the sensor sends

$$\mathbf{s}_k = T\mathbf{y}_k = TC\mathbf{x}_k + T\mathbf{v}_k. \tag{2.3}$$

Note that this requires that all measurements are available to the precoder.

In general, there is no restriction on the number of rows of T, i.e., n_s, and thus the number of packets sent at each sampling instance. By choosing $n_s < n_y$ the number of packets per sampling interval will be reduced. This can be interpreted as some kind of data compression. On the other hand, it is possible to increase the number of packets per sampling interval by choosing $n_s > n_y$, which can be interpreted as adding some redundancy. Obviously, for $n_s = n_y$ the number of packets per sampling interval is unchanged.

Note that there are two special cases of the precoding matrix T, where we do not expect good results:

- T is singular. In this case, there is at least one set $\mathcal{S}$ of transformed measurements, where a subset of $\mathcal{S}$ contains the same information as $\mathcal{S}$. This means that there are transformed measurements that will not improve the estimate of the remote Kalman filter if other transformed measurements are available.

- T is such that the system becomes non-observable if one packet is lost. Obviously, it is also possible to choose T such that the system is not observable even when all packets arrive.

Note that the linear precoding considered in this chapter is a static operation. Obviously, it would also be possible to use a dynamic filter instead, e.g., a Kalman filter, as done in, e.g., Gupta et al. (2009); Xu and Hespanha (2005). However, this requires a sensor with more computational capacities and the ability to store data. Moreover, sending the state estimate instead of the measurements, generally increases the data rate. Due to these drawbacks, this chapter is limited to the simple static linear precoding and we show that significant improvements are possible, even with this relative simple approach.

2.4. Kalman Filtering with Linear Precoding

In this section, we combine Kalman filtering with intermittent observations and linear precoding. Kalman filtering with intermittent observations is considered in Sinopoli et al. (2004); its contribution can be summarized as follows. First, Sinopoli et al. (2004) showed that there exists a critical arrival probability p_c such that the expected value of the error covariance matrix is bounded for $p_c < p$ and also gave a computable upper bound for this critical arrival probability. Moreover, Sinopoli et al. (2004) also gave an upper bound for the expected value of the error covariance matrix. Since the current chapter builds on this work, the main theorems of Sinopoli et al. (2004) are summarized in Chapter A of the appendix.

Combining the precoding (2.3) and the channel model (2.2), we get

$$\mathbf{r}_k = L_k \mathbf{s}_k = L_k TC\mathbf{x}_k + L_k T\mathbf{v}_k \tag{2.4}$$

as input for the remote Kalman filter. Since the time update of the Kalman filter does not depend on the measurements, the time update remains

$$\hat{\mathbf{x}}_{k+1|k} = A\hat{\mathbf{x}}_{k|k}, \tag{2.5a}$$

$$P_{k+1|k} = AP_{k|k}A^\mathsf{T} + W. \tag{2.5b}$$

By replacing C with $L_k TC$ and V with $L_k TVT^\mathsf{T}L_k^\mathsf{T}$ in (A.3), the measurement update becomes

$$\hat{\mathbf{x}}_{k+1|k+1} = \hat{\mathbf{x}}_{k+1|k} + K_k(\mathbf{r}_{k+1} - L_k TC\hat{\mathbf{x}}_{k+1|k}), \tag{2.6a}$$

$$P_{k+1|k+1} = P_{k+1|k} - K_k L_k TCP_{k+1|k}, \tag{2.6b}$$

with

$$K_k := P_{k+1|k}C^\mathsf{T}T^\mathsf{T}L_k^\mathsf{T}\left(L_k TCP_{k+1|k}C^\mathsf{T}T^\mathsf{T}L_k^\mathsf{T} + L_k TVT^\mathsf{T}L_k^\mathsf{T}\right)^{-1}.$$

Remark 2.1. *The term $L_k TCP_{k+1|k}C^\mathsf{T}T^\mathsf{T}L_k^\mathsf{T} + L_k TVT^\mathsf{T}L_k^\mathsf{T}$ can only be inverted when L_k does not contain rows with only zeros. Thus, the erasure matrices are created by deleting a row when the corresponding packet is lost, instead of replacing it by a row with only zeros.*

The Modified Algebraic Riccati Equation (MARE) for this case follows directly from (A.5). By replacing p with $\sum_{e=1}^E w_e(p)$, C with $\tilde{L}_e TC$, and V with $\tilde{L}_e TVT^\mathsf{T}\tilde{L}_e^\mathsf{T}$, we get

$$g_p(X) = AXA^\mathsf{T} + W - \sum_{e=1}^E w_e(p)\Upsilon_e, \tag{2.7}$$

with $\Upsilon_e = AXC^\mathsf{T}T^\mathsf{T}\tilde{L}_e^\mathsf{T}\left(\tilde{L}_e TCXC^\mathsf{T}T^\mathsf{T}\tilde{L}_e^\mathsf{T} + \tilde{L}_e TVT^\mathsf{T}\tilde{L}_e^\mathsf{T}\right)^{-1}\tilde{L}_e TCXA^\mathsf{T}$.

As already stated, Sinopoli et al. (2004) showed that there exists a critical arrival probability p_c such that the expected value of the error covariance matrix is bounded for $p_c < p$, i.e.,

$$\mathrm{E}[P_k] \leq M_{P_0} \ \forall k, \qquad \text{for } p_c < p \leq 1 \text{ and } \forall P_0 \geq 0$$

and unbounded otherwise, see Theorem A.1 for the details. Loosely speaking, the critical arrival probability p_c separates between networks where remote estimation is possible and networks where remote estimation is not possible due to an unbounded estimation error. Since this critical arrival probability can not be calculated directly, Sinopoli et al. (2004) also contains a computable upper bound for the critical arrival probability. Moreover, Sinopoli et al. (2004) showed that the expected value of the error covariance matrix can be bounded by

$$\mathrm{E}[P_k] \leq \bar{P}_k, \qquad \forall\, \mathrm{E}[P_0] \geq 0,$$

where $\bar{P}_k$ is found by the sequence $\bar{P}_{k+1} = g_p(\bar{P}_k)$, $\bar{P}_0 = E[P_0]$. Moreover, $\lim_{k\to\infty} \bar{P}_k = \bar{P}_\infty$ where $\bar{P}_\infty$ is the fixed point of (2.7), i.e., $\bar{P}_\infty = g_p(\bar{P}_\infty)$, see Theorem A.3 for the details. To get the upper bound of the critical arrival probability and the upper bound of the error covariance matrix for the case that linear precoding is used to improve these quantities, Theorem A.2 and A.4 must be adapted. In doing so, we get the following two theorems.

Theorem 2.2. *An upper bound $\bar{p}_c$ of the critical arrival probability is given by the solution of the following optimization problem*

$$\bar{p}_c = \arg\min_p \Psi(Y, Z_1, \ldots, Z_E, T) > 0, \qquad 0 \leq Y \leq I,$$

where

$$\Psi(Y, Z_1, \ldots, Z_E, T) = \begin{bmatrix} Y & \Theta_1 & \Theta_2 & \cdots & \Theta_E \\ \star & Y & 0 & \cdots & 0 \\ \star & \star & Y & & 0 \\ \vdots & \vdots & \vdots & \ddots & \\ \star & \star & \star & & Y \end{bmatrix},$$

with $\Theta_e = \sqrt{w_e(p)}(YA + Z_e \tilde{L}_e TC).$

Theorem 2.3. *If $p > \bar{p}_c$, then the matrix $\bar{P}_\infty = g_p(\bar{P}_\infty)$ is given by*

(a) $\bar{P}_\infty = \lim_{k\to\infty} \bar{P}_k$; $\bar{P}_{k+1} = g_p(\bar{P}_k)$ *where* $\bar{P}_0 \geq 0$.

(b) $\bar{P}_\infty = \arg\max_S \mathrm{Tr}(S)$ *subject to* $\Gamma(S, T) \geq 0$, $S \geq 0$,
  where

$$\Gamma(S, T) = \begin{bmatrix} ASA^\mathsf{T} + W - S & \Pi_1 & \Pi_2 & \cdots & \Pi_E \\ \star & \Xi_1 & 0 & \cdots & 0 \\ \star & \star & \Xi_2 & & 0 \\ \vdots & \vdots & \vdots & \ddots & \\ \star & \star & \star & & \Xi_E \end{bmatrix},$$

with $\Pi_e = \sqrt{w_e(p)}\, ASC^\mathsf{T} T^\mathsf{T} \tilde{L}_e^\mathsf{T}$ *and* $\Xi_e = \tilde{L}_e TCSC^\mathsf{T} T^\mathsf{T} \tilde{L}_e^\mathsf{T} + \tilde{L}_e TV T^\mathsf{T} \tilde{L}_e^\mathsf{T}.$

The proofs of these theorems follow the same line as the one in Liu and Goldsmith (2004b) and are thus omitted.

For a discussion on the tightness of the upper bound $\bar{p}_c$ see Blind and Allgöwer (2014), where the stabilizability of a networked control system with loss and delay is studied. Thereby, the stabilizability is not defined via the convergence of a MARE but the existence of a controller such that the closed loop is mean square stable. Within this framework, the critical arrival probability p_c is defined as the infimum over all arrival probabilities such that there exists a stabilizing controller whereas the upper bound $\bar{p}_c$ is defined as the minimum over all arrival probabilities such that there exists a stabilizing controller.

Now we can choose the precoding matrix T according to the two goals:

Goal 1: Make the Kalman filter more robust against packet loss. Here we choose the precoding matrix T in such a way that the upper bound $\bar{p}_c$ of the critical arrival probability is minimized. Therefore, we use Theorem 2.2 and search for $Y, Z_1, \ldots, Z_E$ and T such that $\arg\min_p \Psi(Y, Z_1, \ldots, Z_E, T) > 0$ is minimized. Note that we have to use the upper bound $\bar{p}_c$ of the critical arrival probability to obtain the optimal precoding matrix T because we cannot calculate the critical arrival probability p_c exactly.

Goal 2: Minimize the estimation error. Here we choose T in such a way that the estimation error $\mathbf{e}_k = \mathbf{x}_k - \hat{\mathbf{x}}_k$ is minimized for a given p. Therefore, we use Theorem 2.3 and search for T such that $\text{Tr}(\bar{P}_\infty)$ is minimized. This approach is especially interesting when the communication system is given and the packet arrival probabilities are known. Again, we have to use the upper bound $\bar{P}_\infty$ because $\lim_{k\to\infty} \text{E}[P_k]$ is not known exactly.

To simplify the presentation of the examples in the next section, we finally show that the rows of the precoding matrix T can be normalized without affecting the precoding.

Lemma 2.4. *Let T be an arbitrary precoding matrix. The solution of the MARE $X = g_p(X)$ and the upper bounds on the critical arrival probability $\bar{p}_c$ are invariant with respect to a scaling of the rows of T, i.e., by replacing T with NT where $N = \text{diag}(n_{11}, \ldots, n_{n_s n_s})$ and $n_{ii} \neq 0$, $\forall i$.*

Proof. A scaling of the rows of the precoding matrix T corresponds to the replacement of $L_k T$ in (2.4) by $L_k NT$. Since L_k contains different row vectors of the identity matrix, a scaling of the columns of L_k by $L_k N$ can also be written as a scaling of the rows of L_k, i.e., $\tilde{N} L_k$, where $\tilde{N} = L_k N L_k^\mathsf{T}$ is a $n_s \times n_s$ invertible diagonal matrix containing n_s diagonal elements of N. Hence, the net effect of a scaling of the rows of T is to scale $\mathbf{r}_k$ which is an invertible process. $\qquad\square$

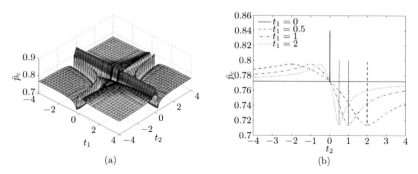

(a) (b)

Figure 2.2.: The upper bound $\bar{p}_c$ of the critical arrival probability for Example 1.[1]

2.5. Examples

We use two examples to show the benefits of precoding the measurements before sending them over a lossy communication system to a remote estimator. Since the aim of this thesis is to improve the control performance by optimizing the communication system, we restrict ourselves to the case $n_s = n_y$, i.e., we do not change the packet rate. Moreover, for the sake of clarity, we consider two relatively simple examples with two measurements ($n_y = 2$). By parameterizing the precoding transform as $T = \begin{bmatrix} 1 & t_1 \\ t_2 & 1 \end{bmatrix}$ and plotting $\bar{p}_c$ or $\mathrm{Tr}(\bar{P}_\infty)$ over t_1 and t_2, we easily see the influence of the precoding matrix T on these performance measures. Note that for this parameterization of T the origin ($t_1 = t_2 = 0$) corresponds to the case without a precoding.

Example 1

First, we reconsider the motivating example from the introduction of this chapter with

$$A = \begin{bmatrix} 2.5 & 0 \\ 0 & 2 \end{bmatrix} \quad \text{and} \quad C = \begin{bmatrix} 1 & 0 \\ 0 & 1 \end{bmatrix}.$$

Moreover, we set $W = 10I$ and $V = 2.5I$.

Without the precoding transform, we obtain $\bar{p}_c = 0.84$. Thus, when more than 84% of the packets arrive the expected value of the covariance matrix P_k is guaranteed to be bounded.

To achieve Goal 1, i.e., make the Kalman filter more robust against packet loss, we search for a transform that minimizes $\bar{p}_c$. To see how $\bar{p}_c$ depends on the precoding matrix T, Figure 2.2 shows $\bar{p}_c$ over the parameters t_1 and t_2. We see that $\bar{p}_c$ can be

---

[1]Note that there should be a straight line of increased $\bar{p}_c$ for $t_1 = 1/t_2$ in Figure 2.2a but due to the rectangular grid it looks more like multiple peaks.

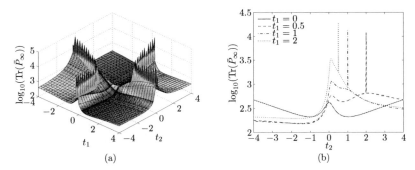

(a)   (b)

Figure 2.3.: The upper bound $\text{Tr}(\bar{P}_\infty)$ of the expected estimation error for Example 1.[2]

significantly reduced by a proper precoding transform and it would be easy to find a transform such that $\bar{p}_c < 0.75$. On the other hand, note that $\bar{p}_c$ is dramatically increased along the line $t_2 = 0$. This is due to the fact that for $t_2 = 0$ the more unstable first mode is not observable when $s_{k,1}$ is lost.

Not surprisingly, $\bar{p}_c$ is also increased along the line $t_1 = 1/t_2$ where T is singular. Interestingly, $\bar{p}_c$ is relatively small when T is close to singular. This effect is also observed in the field of MDC and can be explained as follows. The key of the choice of the precoding matrix T is that each row of TC has to point into the most important direction for the state estimation task but also to make each row a little bit different so that the combination of different rows allows a better estimation of the state. This is analog to MDC where the most important direction is the direction of maximum scatter of the data vectors which is called the principal component, see Goyal (2001).

In order to achieve Goal 2, i.e., minimize the estimation error, we ask for a transform that minimizes $\text{Tr}(\bar{P}_\infty)$ for a fixed p. Here, we choose $p = 0.85$ and get $\text{Tr}(\bar{P}_\infty) = 415.94$ without the precoding. Using MATLAB's `fminsearch` function gives us $T_{\text{opt}} = \begin{bmatrix} 1 & 0.6803 \\ -1.4813 & 1 \end{bmatrix}$ and $\text{Tr}(\bar{P}_\infty) = 148.69$, which improves the quality of the Kalman filter considerably. However, since the corresponding optimization problem is nonconvex we can not guarantee that T_{opt} is globally optimal. On the other hand, by a bad choice of the precoding matrix T, the estimation error will be increased. This can be seen in Figure 2.3, which shows how $\log_{10}(\text{Tr}(\bar{P}_\infty))$ depends on the precoding matrix T. Here, $\text{Tr}(\bar{P}_\infty)$ is huge when T is singular, i.e., along the line $t_1 = 1/t_2$.

Interestingly, in this example, there is a tradeoff between robustness against packet loss and the quality of the remote estimates for a given arrival probability. This is best seen when comparing Figure 2.2b and 2.3b. When choosing a transformation T that gives a relatively high robustness against packet loss the resulting trace of the error

---

[2]Note that there should be a straight line of increased $\text{Tr}(\bar{P}_\infty)$ for $t_1 = 1/t_2$ in Figure 2.3a but due to the rectangular grid it looks more like multiple peaks.

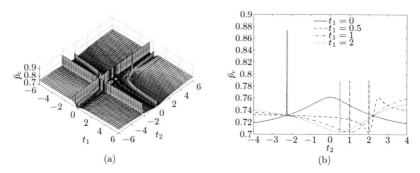

(a)   (b)

Figure 2.4.: The upper bound $\bar{p}_c$ of the critical arrival probability Example 2.[3]

covariance matrix will be relatively large. E.g., when choosing t_2 positive, $\bar{p}_c$ will be relatively small but $\text{Tr}(\bar{P}_\infty)$ relatively large.

Example 2

The previous example was composed of two unstable systems which became non-observable when one of the packets was lost. Now, we look at a system which is observable if any of the two packets arrive:

$$A = \begin{bmatrix} 2.5 & 0.25 \\ 1 & 2 \end{bmatrix} \quad \text{and} \quad C = \begin{bmatrix} 1 & 0 \\ 1 & 1 \end{bmatrix}.$$

As in the previous example, we set $W = 10I$ and $V = 2.5I$. This system has two unstable eigenvalues $\lambda_1 \approx 2.809$ and $\lambda_2 \approx 1.691$. Note that this system is not observable for $t_1 = \pm\tilde{t}_1$, $\tilde{t}_1 = 1/\sqrt{5} \approx 0.447$ or $t_2 = \pm\tilde{t}_2$, $\tilde{t}_2 = \sqrt{5} \approx 2.236$ when only the corresponding packet arrives. More precisely, for $t_1 = -\tilde{t}_1$ the eigenmode λ_1 becomes non-observable when only the first packet $s_{k,1}$ arrives; for $t_1 = +\tilde{t}_1$ the eigenmode λ_2 becomes non-observable when only the first packet $s_{k,1}$ arrives. Similarly, for $t_2 = -\tilde{t}_2$ the eigenmode λ_1 becomes non-observable when only the second packet $s_{k,2}$ arrives; for $t_2 = +\tilde{t}_2$ the eigenmode λ_2 becomes non-observable when only the second packet $s_{k,2}$ arrives.

Figure 2.4 shows how $\bar{p}_c$ depends on T and we see similar effects as in the previous example for the two special cases of T. Interestingly, $\bar{p}_c$ is only increased for $t_1 = -\tilde{t}_1$ but not for $t_1 = +\tilde{t}_1$ although in both cases the system is not observable when only the first packet $s_{k,1}$ arrives. This is due to the fact that for $t_1 = -\tilde{t}_1$ the more unstable

---

[3]Note that there should be a straight line of increased $\bar{p}_c$ for $t_1 = 1/t_2$ in Figure 2.4a but due to the rectangular grid it looks more like multiple peaks.

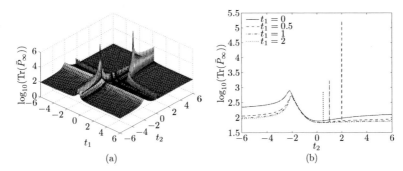

Figure 2.5.: The upper bound $\text{Tr}(\bar{P}_\infty)$ of the expected estimation error for Example 2.[4]

eigenmode λ_1 becomes non-observable and for $t_1 = +\tilde{t}_1$ the less unstable eigenmode λ_2 becomes non-observable. The same effect can be observed for t_2.

Without the precoding we get $\bar{p}_c = 0.7618$. For the transform that minimizes $\bar{p}_c$ MATLAB's `fminsearch` gives us $T_{\text{opt}} = \begin{bmatrix} 1 & 17.2102 \\ -0.053 & 1 \end{bmatrix}$. With this precoding matrix, we get $\bar{p}_c = 0.7039$. Hence, even if the original system is observable when only one packet arrives, $\bar{p}_c$ can still be reduced significantly by precoding the measurements.

As in the previous example, we now fix p and search for the transform that minimizes $\text{Tr}(\bar{P}_\infty)$. In this example, we set p to 0.88. Figure 2.5 shows how $\log_{10}(\text{Tr}(\bar{P}_\infty))$ depends on the precoding matrix T. Again, we see a significant influence of the two special cases. In this example, we get $\text{Tr}(\bar{P}_\infty) = 103.27$ for the original system and $\text{Tr}(\bar{P}_\infty) = 82.76$ for the optimal transform $T_{\text{opt}} = \begin{bmatrix} 1 & 4.1512 \\ 0.5653 & 1 \end{bmatrix}$. Hence, the linear precoding improves the quality of the Kalman filter although the original system is observable when only one packet arrives.

As already stated, the system is not observable for $t_1 = \pm\tilde{t}_1$ or $t_2 = \pm\tilde{t}_2$ when only the corresponding packet arrives. Moreover, the precoding matrix T is singular when choosing $t_1 = +\tilde{t}_1$ and $t_2 = +\tilde{t}_2$ or $t_1 = -\tilde{t}_1$ and $t_2 = -\tilde{t}_2$. This explains the huge values of $\bar{p}_c$ and $\text{Tr}(\bar{P}_\infty)$, for these values of t_1 and t_2 in Figure 2.4a and 2.5a.

2.6. Summary

In this chapter, we showed how to design a linear precoding matrix for a remote Kalman filter in the presence of lossy links. We showed that the precoding matrix can be chosen such that the remote Kalman filter will be more robust against packet loss or chosen such that the estimation error is reduced.

---

[4]Note that there should be a straight line of increased $\text{Tr}(\bar{P}_\infty)$ for $t_1 = 1/t_2$ in Figure 2.5a but due to the rectangular grid it looks more like multiple peaks.

Chapter 3.

Optimization within the Transport Layer: Retransmitting Measurements

In the previous chapter, we used a method located within the presentation layer of the ISO/OSI Stack, called linear precoding, to optimize remote estimation over a communication system with random packet loss. In this chapter, we study the same problem but assume that we have some more control over the communication system and can affect how measurement packets are handled within the transport layer of the communication system: The receiver acknowledges each successfully arrived measurement packet and the sender adds all measurement packets, that are not yet acknowledged, to the next measurement packet. Although previous works found no benefit in acknowledging and retransmitting measurement packets, we show that the presented retransmission scheme results in optimal estimates. In contrast to other approaches that give optimal estimates, the presented retransmission scheme does not require huge data transfers or a preprocessing of the measurements by the sensor. This chapter is based on Blind and Allgöwer (2013a)

3.1. Problem Setup

As in the previous chapter, the considered problem setup is similar to the one of Sinopoli et al. (2004) and depicted in Figure 3.1. The goal is to remotely estimate the state of a linear system

$$\mathbf{x}_{k+1} = A\mathbf{x}_k + \mathbf{w}_k \tag{3.1a}$$

$$\mathbf{y}_k = C\mathbf{x}_k + \mathbf{v}_k, \tag{3.1b}$$

where $\mathbf{x}_k \in \mathbb{R}^{n_x}$ is the state, $\mathbf{y}_k \in \mathbb{R}^{n_y}$ the measurement, $\mathbf{w}_k \in \mathbb{R}^{n_x}$ and $\mathbf{v}_k \in \mathbb{R}^{n_y}$ are Gaussian random vectors with zero mean and covariance matrices $W \in \mathbb{R}^{n_x \times n_x}$, $W \geq 0$, and $V \in \mathbb{R}^{n_y \times n_y}$, $V > 0$, respectively. Moreover, $\mathbf{w}_k$ and $\mathbf{w}_s$ are independent for $s \neq k$, the pair (A, C) is detectable, and $(A, W^{\frac{1}{2}})$ is stabilizable.

In order to remotely estimate the state, the sensor sends a packet $\mathbf{s}_k$ at time k over a communication system, which randomly drops packets. As will be discussed in the next section, there is some degree of freedom in what to send. Thus, we do not fix the

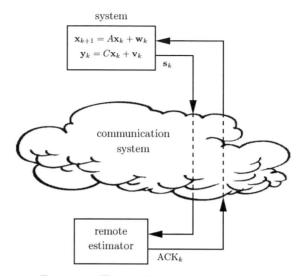

Figure 3.1.: The considered problem setup.

content of these packets yet. The remote estimator receives

$$\mathbf{r}_k = \begin{cases} \mathbf{s}_k & \text{arrival,} \\ \emptyset & \text{loss.} \end{cases}$$

We use l to denote the time when the last packet was received by the remote estimator. Moreover, we use the superscripts s and e to distinguish between the data at the sensor and the one at the estimator. The information set of the sensor is the set of all measurements, i.e.,

$$\mathcal{I}_k^{\mathrm{s}} := \{\mathbf{y}_0, \dots, \mathbf{y}_k\};$$

the information set of the remote estimator is the set of all received packets, i.e.,

$$\mathcal{I}_k^{\mathrm{e}} := \{\mathbf{r}_0, \dots, \mathbf{r}_k\}.$$

Note that the state estimate and the error covariance matrix of the sensor and the one of the remote estimator just depends on the different information sets, i.e.,

$$\hat{\mathbf{x}}_{k|k}^{\xi} := \mathrm{E}\big[\mathbf{x}_k | \mathcal{I}_k^{\xi}\big],$$
$$P_{k|k}^{\xi} := \mathrm{E}\big[\big(\mathbf{x}_k - \hat{\mathbf{x}}_{k|k}^{\xi}\big)\big(\mathbf{x}_k - \hat{\mathbf{x}}_{k|k}^{\xi}\big)^T\big],$$
$$\hat{\mathbf{x}}_{k+1|k}^{\xi} := \mathrm{E}\big[\mathbf{x}_{k+1} | \mathcal{I}_k^{\xi}\big],$$
$$P_{k+1|k}^{\xi} := \mathrm{E}\big[\big(\mathbf{x}_{k+1} - \hat{\mathbf{x}}_{k+1|k}^{\xi}\big)\big(\mathbf{x}_{k+1} - \hat{\mathbf{x}}_{k+1|k}^{\xi}\big)^T\big],$$

where $\xi \in \{\mathrm{s}, \mathrm{e}\}$.

3.2. Transmission and Retransmission Schemes

3.2.1. Previous Transmission and Retransmission Schemes

In the original problem formulation of Sinopoli et al. (2004), only the current measurement $\mathbf{y}_k$ is sent, i.e., $\mathbf{s}_k = \mathbf{y}_k$. Clearly, sending only the current measurement $\mathbf{y}_k$ is not optimal and thus different approaches to reduce the estimation error are proposed in recent literature. Figure 3.2 depicts these transmission and retransmission schemes. When the sensor has enough computational resources, it can calculate the state estimate $\hat{\mathbf{x}}_k^s$ and then send it to the remote estimator. Another approach is the addition of some, or all, previous measurements to the current measurement packet.

In Xu and Hespanha (2005), the remote estimation of a continuous-time system is considered. It is proposed to use a smart sensor with enough computational ability to calculate the state estimate $\hat{\mathbf{x}}_k$ and send it over the network. In Gupta et al. (2009), it is shown that the optimal state estimate $\hat{\mathbf{x}}_k^{e,*}$ is

$$\hat{\mathbf{x}}_k^{e,*} = \mathrm{E}[\mathbf{x}_k | \mathbf{y}_0, \ldots, \mathbf{y}_l], \tag{3.2}$$

i.e., it is calculated based on all measurements up to time l. Furthermore, Gupta et al. (2009) presents two methods to achieve optimal estimates. In the first approach, depicted in Figure 3.2b, the sensor sends its full information set, i.e., $\mathbf{s}_k = \mathcal{I}_k^s$. In the second approach, depicted in Figure 3.2c, the sensor calculates the state estimate and sends it to the remote estimator, i.e., $\mathbf{s}_k = \hat{\mathbf{x}}_k^s$. Unfortunately, both approaches might be difficult, or even impossible, to realize in practice. Sending the full information set $\mathcal{I}_k^s$ is impossible because the packets become too large to be transmitted. Calculating the state estimate at the sensor is only possible when the sensor has enough computational resources and knowledge of the system. With the help of an observer based estimator and always sending the current and several previous measurements, i.e., $\mathbf{s}_k = \{\mathbf{y}_{k-p}, \ldots, \mathbf{y}_k\}$, Epstein et al. (2008); Shi et al. (2010) show that the error covariance is bounded with a high probability, i.e., $\Pr\{P_k \leq M\} \geq 1 - \epsilon$. This approach is depicted in Figure 3.2d. Note that all these approaches require to increase the payload of a measurement packet. However, except for the case of sending the full information set, this is not a problem for many communication systems, where the overhead due to the headers is large when compared with the size of a measurement. In this case, the more important factor is the packet rate and not the size of the payload.

Another approach potentially improving the remote estimates is the retransmission of lost measurements, as considered in Leong et al. (2008), where the measurement $\mathbf{y}_k$ is sent, and Gupta (2010), where the state estimate $\hat{\mathbf{x}}_k^s$ is sent. These approaches are depicted in Figure 3.2e and 3.2f. In both works, the lost packet is retransmitted until the corresponding acknowledgement is received but the following packets are dropped since the packet rate or payload size is not increased. Both conclude that this retransmission scheme does not improve the state estimates.

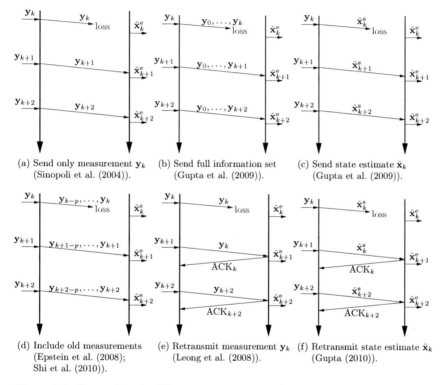

(a) Send only measurement $\mathbf{y}_k$ (Sinopoli et al. (2004)).

(b) Send full information set (Gupta et al. (2009)).

(c) Send state estimate $\hat{\mathbf{x}}_k$ (Gupta et al. (2009)).

(d) Include old measurements (Epstein et al. (2008); Shi et al. (2010)).

(e) Retransmit measurement $\mathbf{y}_k$ (Leong et al. (2008)).

(f) Retransmit state estimate $\hat{\mathbf{x}}_k$ (Gupta (2010)).

Figure 3.2.: Comparing the different transmission and retransmission schemes found in recent literature.

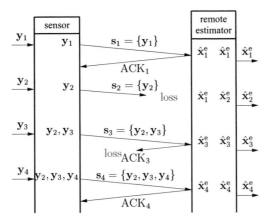

Figure 3.3.: An example of the proposed retransmission scheme.

3.2.2. New Retransmission Scheme

Since the previously published transmission and retransmission schemes are either not optimal, require to send all measurements, or require to calculate the state estimate at the sensor, we present yet another approach. Similar to Epstein et al. (2008); Shi et al. (2010) we add previous measurements to the current measurement packet. However, instead of adding a fixed number of old measurements, we add all non-acknowledged measurements. When receiving the measurement $\mathbf{y}_k$, the remote estimator sends an acknowledgement ACK_k back to the sensor to acknowledge the reception of $\mathbf{y}_k$ and all previous measurements. When the sensor has one or more non-acknowledged measurements, it adds these measurements to the next measurement packet. Thus, at time k the sensor sends

$$\mathbf{s}_k = \{\mathbf{y}_{l+1}, \ldots, \mathbf{y}_k\}. \tag{3.3}$$

In doing so, the remote estimator must be modified as follows. In addition to storing the estimate of the current state $\hat{\mathbf{x}}_k^e$, the estimator must also keep the state estimate, which was calculated when the last measurement packet was received, i.e., $\hat{\mathbf{x}}_l^e$. Similar to all previously discussed schemes, the estimator can not perform the measurement update when the current packet is lost, i.e., when $\mathbf{r}_k = \emptyset$. Upon the reception of a measurement packet, i.e., when $\mathbf{r}_k = \mathbf{s}_k = \{\mathbf{y}_{l+1}, \ldots, \mathbf{y}_k\}$, the estimator has to recalculate the state estimates $\hat{\mathbf{x}}_{l+1}^e, \ldots, \hat{\mathbf{x}}_{k-1}^e$. Since this measurement packet contains all missing measurements, the Kalman filter can now perform the measurement updates while recalculating the state estimate.

Figure 3.3 depicts an example of the proposed retransmission scheme. We assume that the first packet contains only the first measurement $\mathbf{y}_1$. Since it is successfully received, the remote estimator can perform the time and measurement update. Unfortunately, the second measurement packet is lost. Thus, the remote estimator keeps $\hat{\mathbf{x}}_1^e$ but also performs a time update to get $\hat{\mathbf{x}}_2^e$. Since the sensor does not receive an acknowledgment that indicates the reception of $\mathbf{y}_2$, it adds $\mathbf{y}_2$ to the next measurement packet and sends $\mathbf{s}_3 = \{\mathbf{y}_2, \mathbf{y}_3\}$. Since this packet is successfully received, the remote estimator recalculates the state estimate based on the state estimate when the last measurement packet was received, i.e., $\hat{\mathbf{x}}_1^e$, and the measurements $\mathbf{y}_2$ and $\mathbf{y}_3$ to get $\hat{\mathbf{x}}_3^e$. However, the corresponding acknowledgment ACK_3 is lost. Since the two measurements $\mathbf{y}_2$ and $\mathbf{y}_3$ are not yet acknowledged, the sensor adds them to the next packet and sends $\mathbf{s}_4 = \{\mathbf{y}_2, \mathbf{y}_3, \mathbf{y}_4\}$, which is successfully received. Thus, the remote estimator performs a measurement update on $\hat{\mathbf{x}}_3^e$ to get $\hat{\mathbf{x}}_4^e$.

After the presentation of the proposed retransmission scheme, we now analyze this approach and show that this approach gives optimal estimates. Moreover, we also derive the probability mass function of the number of measurements per packet, their expected value, the probability mass function of the error covariance matrix, and its expected value. For simplicity, we thereby assume that packets either arrive on time or are lost.

Theorem 3.1. *The proposed retransmission scheme, gives optimal estimates as defined in (3.2).*

Proof. By sending an acknowledgement ACK_l, the receiver acknowledges the reception of all measurements up to time l, i.e., it acknowledges that its information set is $\mathcal{I}_l^e = \{\mathbf{y}_0, \mathbf{y}_1, \dots, \mathbf{y}_l\}$. Thus, at time $k > l$, it is sufficient to send only the difference between the information set of the sensor and the information set of the estimator, i.e., $\mathbf{s}_k = \mathcal{I}_k^d := \mathcal{I}_k^s \setminus \mathcal{I}_l^e = \{\mathbf{y}_{l+1}, \dots, \mathbf{y}_k\}$. $\qquad\square$

As already stated, adding non-acknowledged measurements to the current measurement packet might be problematic. This will be the case when the CAN (Controller Area Network) protocol is used for packet transmission since its maximal payload is only 8 byte. However an Ethernet packet is at least 64 byte long (or even 84 byte, when taking the preamble and interframe gap into account) and has a minimum payload size of 42 byte. In this case, adding non-acknowledged measurements to the current measurement packet should be no problem. Here, when worrying about the usage of resources, only the packet rate, i.e., the number of packets per time, is of interest, not the number of measurements contained in one packet. Thus, whether adding non-acknowledged measurements to the current measurement packet is problematic or not, depends on the communication protocol.

The following statements give the probability mass function of the number of measurements per packet, the probability that more than M measurements are transmitted in one packet, and the expected number of measurements per packet.

Lemma 3.2. *Suppose packet loss is independent and identically distributed (iid); the arrival probability of measurement packets is p_y and the arrival probability of acknowl-*

edgement packets is p_{ACK}. Then, the Probability Mass Function (PMF) of measurements per packet N_y is

$$\Pr\{N_y = i\} = p_y p_{\text{ACK}} (1 - p_y p_{\text{ACK}})^{i-1}, \quad i \geq 1$$

Proof. This lemma follows by noting that we have to transmit i measurements when in all $i - 1$ previous time steps either the measurement or the acknowledgement packet was lost and at the last but i-th time steps both arrived. $\square$

Corollary 3.3. *The probability that more than M measurements are transmitted in one measurement packet is*

$$\Pr\{N_y > M\} = (1 - p_y p_{\text{ACK}})^M.$$

Corollary 3.4. *The expected number of measurements per packet is*

$$\text{E}[N_y] = 1/p_y p_{\text{ACK}}.$$

When considering the bit rate, $\text{E}[N_y]$ is the factor how much the payload is increased when compared to the case that each packet contains only one measurement. Thus, when $\text{E}[N_y]n_y < n_x$, then the expected bit rate between sensor and remote estimator required by the proposed retransmission scheme is less than the bit rate that is required when the state estimate is sent at each time.

Remark 3.5. *The number of measurements that must be stored by the sensor is equal to the number of measurements per packet.*

Now, we study the properties of the error covariance matrix. Therefore, we use

$$g(X) := AXA^{\mathsf{T}} + W$$

to denote the effect of a time update on the error covariance matrix. Moreover, $g^i(X)$ means that g is applied i times, i.e., $g^i(X) = g(g^{i-1}(X))$ with the definition $g^0(X) := X$. Furthermore, $\bar{P}$ is the solution of the algebraic Riccati equation

$$\bar{P} = A\bar{P}A^{\mathsf{T}} + W - A\bar{P}C^{\mathsf{T}}(C\bar{P}C^{\mathsf{T}} + V)^{-1}C\bar{P}A^{\mathsf{T}}.$$

Theorem 3.6. *Suppose packet loss is independent and identically distributed (iid); the arrival probability of measurement packets is p_y. Moreover, assume that $P_{0|0} = \bar{P}$. Then, the Probability Mass Function (PMF) of the error covariance matrix is*

$$\Pr\{P_{k|k} = g^i(\bar{P})\} = p_y(1 - p_y)^i, \quad i \geq 0.$$

Proof. Remember that the estimator has the full information set whenever a measurement packet arrives. In this case, the error covariance will be exactly the one of the standard Kalman filter without packet losses, i.e., $P_{k|k} = \bar{P}$. Since this happens with probability p_y, we already proved Theorem 3.6 for $i = 0$. Now, we consider the case $i > 0$. Here, the probability that all previous i measurement packets were lost and the last but $(i + 1)$-th was received is $p_y(1 - p_y)^i$. In this case, i time updates were necessary and thus the error covariance will be $g^i(\bar{P})$. $\square$

Remark 3.7. *The assumption $P_{0|0} = \bar{P}$ is only necessary to keep the theorem and proof simple. Since the standard Kalman filter without packet losses converges to $\bar{P}$ for all $P_{0|0}$ and the fact that we get the same error covariance matrix whenever a measurement packet arrives, this assumption is not restrictive for practical applications.*

Corollary 3.8. *The expected error covariance is*

$$\mathrm{E}[P_{k|k}] = \sum_{i=0}^{\infty} g^i(\bar{P}) p_y (1 - p_y)^i.$$

Remark 3.9. *Lost acknowledgements affect only the number of measurements per packet but not the properties of the remote estimator.*

Note that an exact mathematical analysis becomes much more complex when the packet size is limited. When the packet size is not limited, we have the full information set whenever a measurement packet arrives. Thus the error covariance will be $\bar{P}$ after the arrival of a measurement packet. Obviously, this is not the case when measurements are missing due to a limited packet size. However, from a practical point of view, this is not critical. As already stated, the size of the payload is large for most protocols. Thus, the probability that not all missing measurements fit into a packet of a reasonable size is very small. Moreover, in this case, a good state estimate can be obtained by multiplying the inverse of the observability matrix with the vector of recent measurements. As suggested in Epstein et al. (2008), this state estimate can then be improved by using a Kalman filter for the remaining measurements. In doing so, it is possible to guarantee that the error covariance is bounded after the arrival of a measurement packet.

3.3. Example

In this section, we demonstrate the benefit of the proposed retransmission scheme with the help of a simple example. Therefore, we consider the following system.

$$\mathbf{x}_{k+1} = \begin{bmatrix} 1.4 & 0 \\ 0.1 & 1.2 \end{bmatrix} \mathbf{x}_k + \begin{bmatrix} 1 \\ 0 \end{bmatrix} u_k + \mathbf{w}_k$$

$$y_k = \begin{bmatrix} 0 & 1 \end{bmatrix} \mathbf{x}_k + v_k,$$

with $W = I$, and $V = 1$. We assume that measurement and acknowledgement packets arrive with a probability of $p_y = p_{\mathrm{ACK}} = 0.6$. The loop is closed by the controller

$$u_k = F\hat{\mathbf{x}}_k,$$

where the control gain F is chosen such that the poles of the closed loop system are 0.4 and 0.5. Finally, we assume that all control packets arrive.

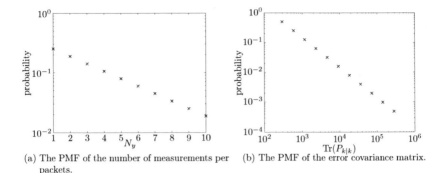

(a) The PMF of the number of measurements per packets.

(b) The PMF of the error covariance matrix.

Figure 3.4.: The Probability Mass Function (PMF) of the measurements per packet and the error covariance matrix.

Figure 3.4 shows the probability mass function of the measurements per packet and the error covariance matrix. From Figure 3.4a and Lemma 3.2, we see that the probability that i measurements are contained in one packet decays exponentially with i. Figure 3.4b depicts the probability mass function of the error covariance matrix. Since only a prediction step is performed whenever a packet is lost, the error covariance matrix grows exponentially with the number of consecutive losses. At the same time, the probability that i consecutive packets were lost, decays exponentially with i.

A comparison of the expected error covariance matrix of the proposed retransmission scheme with the lower and upper bound of the expected error covariance matrix for the case that only the current measurement $\mathbf{y}_k$ is sent is depicted in Figure 3.5. Not surprisingly, the expected value of the error covariance matrix is always smaller than the upper bound of the error covariance matrix for the case that only the current measurement is sent. Nevertheless, the expected value of the error covariance matrix of the proposed retransmission scheme remains larger than the lower bound for the case that only the current measurement is sent.

Finally, Figure 3.6 shows the mean of the estimation error and the normed state for the proposed scheme with retransmissions and the case that only the current measurement $\mathbf{y}_k$ is sent, obtained from 10000 simulations of the closed loop. As already shown, we get a smaller estimation error when lost measurements are retransmitted within the next measurement packet. Remember that the state estimate and thus also the control input is calculated without the correction step when a measurement packet is lost since the retransmitted measurement is part of the next measurement packet.

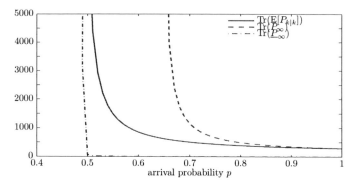

Figure 3.5.: The expected value of the error covariance matrix as well as the upper and lower bound of the expected error covariance matrix for the case that only the current measurement $\mathbf{y}_k$ is sent.

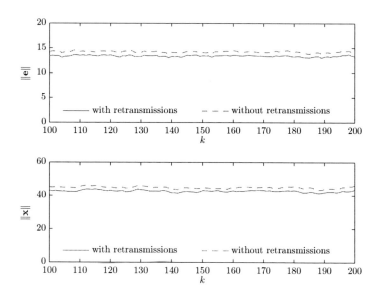

Figure 3.6.: The estimation error and the normed state with and without retransmissions.

3.4. Summary

We presented an approach to effectively retransmit lost measurements. To keep the packet rate and sampling time unchanged, we suggest that the remote estimator acknowledges successfully received measurements and the sensor adds all unacknowledged measurements to the next measurement packet. When the remote estimator does not receive a measurement it just performs the time update step. When a measurement packet with missing measurements is received, the remote estimator recalculates all the old state estimates, where only the time update was performed so far. In doing so, the state estimates of the remote estimator are optimal estimates as defined in Gupta et al. (2009). Finally, we also derived important properties of the error covariance matrix and the number of measurements per packet.

Chapter 4.

Optimization within the Transport Layer: Optimal Sampling Time

In the previous chapter, we showed that remote estimates can be improved by retransmitting lost measurements. We assumed that a discrete-time system is given and lost measurements are retransmitted by piggybacking them to the current measurement packet. With this approach, previously lost measurements are outdated when they finally arrive. Within this chapter, we present another approach to improve the control performance by optimizing the transport layer and suggest to increase the sampling time to allow several transmissions within each sampling interval. Therefore, we drop the assumption that a discrete-time system is given and consider the optimal control of a continuous-time system. Since the packets must be generated at certain times, the continuous-time system must be sampled. In doing so, the original problem can be reformulated as the optimal control of a discrete-time system over a packet based communication system, which has been studied thoroughly, but we have the additional freedom to choose the sampling time. It turns out that this choice is not a simple task. Most communication systems will be more reliable when the network load is low, i.e., when the sampling time is long, see Section 4.3 of this chapter but also Chapter 6 for a more lengthy discussion. On the other hand, when considering only the control performance, a short sampling time should be preferred. Due to this tradeoff, the optimal choice of the sampling time is a challenging problem. This chapter is based on Blind and Allgöwer (2013d); an earlier version is published in Blind and Allgöwer (2012b).

4.1. Introduction

In the ISO/OSI reference model, the *transport layer* is responsible for the end-to-end connection between two users. Here, increasing the number of packets is a well known approach to increase the reliability of a communication system. This can either be done by sending each packet multiple times or by retransmitting lost packets. Both approaches require to increase the number of packets per sampling interval, which can be achieved by either increasing the packet rate or by increasing the sampling time.

In Mesquita et al. (2009, 2012); Nair et al. (2010), the sampling time is kept constant and the number of transmitted packets is increased after a loss. Such a proceeding might overload the communication system, resulting in even more loss, see, e.g., Afanasyev et al. (2010); Jacobson (1988); Rom and Sidi (1990); Tanenbaum (2003). Unfortunately, this problem is not addressed in Mesquita et al. (2009, 2012); Nair et al. (2010). Moreover, in most real communication systems, the assumption of independent losses does not hold for packets that are sent at almost the same time. In most cases, these packets will be transmitted in the same super-packet of a lower layer of the ISO/OSI model and/or share the same queues and physical environment, which are the underlying reason of packet loss. Thus, it is very likely that packets that are sent at the same time will suffer the same fate. To sum up, increasing the packet rate might not solve the problem and might make things even worse. Hence, we keep the packet rate unchanged and accept a longer sampling time.

Unfortunately, increasing the sampling time generally reduces the performance of the closed loop system, see, e.g., Levis et al. (1971). Thus, we raise the question whether it is worth to increase the sampling time to increase the arrival probability. To answer this question, we consider the design of the transport layer for networked control systems. Therefore, we introduce two transport layer protocols, where the arrival probability is increased by allowing multiple transmissions per sampling interval. Without increasing the packet rate, this requires to increase the sampling time. Based on these two transport layer protocols, we introduce and analyze four different configurations of the transport layer for networked control systems. To compare these four configurations, we derive analytical expressions for the dependency of the arrival probability and the minimal sampling time on the number of transmissions per sampling interval. With the help of these expressions, we can finally compare the achieved control performance.

The derivation of the control performance is based on the well studied problem setup of optimal control with unreliable communication links. In Imer et al. (2006); Schenato et al. (2007), the optimal control over a communication system with lossy links is studied. Thereby, the authors distinguish between protocols with reliable acknowledgments, called TCP-like protocol, and protocols without acknowledgments, called UDP-like protocol. These works are extended to unreliable acknowledgments in Garone et al. (2008); Kögel (2009); Kögel, Blind, and Allgöwer (2010). In Kögel (2009); Kögel, Blind, and Allgöwer (2010), three different acknowledgment models are considered: only positive acknowledgments (ACK), only negative acknowledgments (NAK), and the case that both (ACKs and NAKs) are sent. In contrast, Garone et al. (2008) considers only the case that ACKs and NAKs are sent.

A similar problem as the one considered in this chapter is studied in Demirel et al. (2011). However, there are some important differences. In Demirel et al. (2011) it is assumed that only measurement packets are lost, whereas we assume that all packets, i.e., control, measurement, and acknowledgment packets, can get lost. Moreover, Demirel et al. (2011) considers the co-design of controllers and transmission schedules in multi-hop *Wireless*HART. Thus, a relatively complex model of the communication

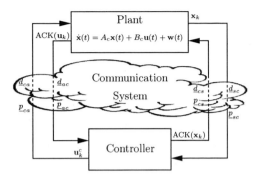

Figure 4.1.: The Networked Control System.

system is used. In contrast, we model each communication channel by a Bernoulli loss process, as done in Garone et al. (2008); Imer et al. (2006); Kögel (2009); Kögel, Blind, and Allgöwer (2010); Schenato et al. (2007); Sinopoli et al. (2004).

4.2. Problem Setup

Figure 4.1 depicts the considered problem setup. A continuous-time system is controlled over a packet based communication system with loss and delay.

4.2.1. Control System

The considered continuous-time system is given by

$$\dot{\mathbf{x}}(t) = A_c\mathbf{x}(t) + B_c\mathbf{u}(t) + \mathbf{w}(t), \qquad \mathbf{x}(0) = \mathbf{x}_0, \tag{4.1}$$

where $\mathbf{x} \in \mathbb{R}^{n_x}$ is the state of the system, $\mathbf{u} \in \mathbb{R}^{n_u}$ the control input applied to the plant, $\mathbf{w} \in \mathbb{R}^{n_x}$ the process noise, and $\mathbf{x}_0$ the initial condition. The process noise $\mathbf{w}$ is assumed to be Gaussian white noise with zero mean and covariance $W_c \in \mathbb{R}^{n_x \times n_x}$. The initial condition $\mathbf{x}_0$ is Gaussian distributed with mean $\bar{\mathbf{x}}_0$ and covariance $X_0 \in \mathbb{R}^{n_x \times n_x}$ and independent of the process noise. Finally, we assume that the state $\mathbf{x}$ can be measured.

This system should be controlled such that the cost

$$J = \mathbf{x}^\mathsf{T}(T_\mathrm{F})F_c\mathbf{x}(T_\mathrm{F}) + \int_0^{T_\mathrm{F}} \mathbf{x}^\mathsf{T}(t)Q_c\mathbf{x}(t) + \mathbf{u}^\mathsf{T}(t)R_c\mathbf{u}(t)\mathrm{d}t \tag{4.2}$$

is minimized.

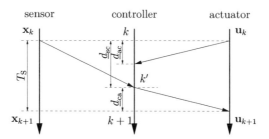

Figure 4.2.: Standard timing with one control, one measurement, and one acknowledgement packet per sampling interval.

4.2.2. Communication System

As depicted in Figure 4.1, a packet based communication system is used to close the loop. All links are modeled by a constant delay and a Bernoulli loss process. Thereby, $\underline{p}_{sc}$ denotes the packet arrival probability from sensor to controller, $\underline{p}_{cs}$ the packet arrival probability from controller to sensor. Similarly, $\underline{p}_{ca}$ and $\underline{p}_{ac}$ are the packet arrival probabilities from controller to actuator and actuator to controller, respectively. The same notation is used for the delays $\underline{d}_{sc}$, $\underline{d}_{cs}$, $\underline{d}_{ca}$, and $\underline{d}_{ac}$. Figure 4.2 depicts the details of the timing, for the case that one control, one measurement, and one acknowledgement packet is sent during each sampling interval. Note that the sensor and actuator are assumed to be synchronized.

4.2.3. Discretization

Since all packets are transmitted over a packet based communication system, the continuous-time system (4.1) is sampled with a constant sampling time T_S. Within this chapter, we assume that the transmission times are the dominating delay and thus determine the sampling time. Figure 4.2 depicts how the sampling time follows from the transmission times. Remember that the sensor and actuator are assumed to be synchronized. To calculate the next control input, the controller waits for the measurement packet and the acknowledgement of the previous control packet. Obviously, when one of these packets is lost, the controller must not wait forever but only a limited time; $\underline{d}_{sc}$ for a measurement packet and $\underline{d}_{ac}$ for an acknowledgement packet. Thus, the next control input is calculated at time $k' = kT_S + \max\{\underline{d}_{sc}, \underline{d}_{ac}\}$. The new control input is then transmitted to the actuator, where it is applied at time $kT_S + \max\{\underline{d}_{sc}, \underline{d}_{ac}\} + \underline{d}_{ca}$. Thus, the minimal sampling time is

$$T_{S,\min} = \max\{\underline{d}_{sc}, \underline{d}_{ac}\} + \underline{d}_{ca}. \tag{4.3}$$

Note that this nevertheless requires a predictive controller. Obviously, another approach would be to choose a faster sampling time and take the delay into account.

To discretize the system, we define $\mathbf{x}_k := \mathbf{x}(kT_S)$ as the state of the corresponding discrete-time system. The input $\mathbf{u}$ is kept constant during each sampling interval, i.e.,

$$\mathbf{u}(t) = \mathbf{u}_k \qquad \text{for } kT_S \leq t < (k+1)T_S.$$

In Levis et al. (1971), the effect of the sampling time on the optimal performance is studied. Therefore, the continuous-time system and its cost are discretized as follows. The discrete-time system evolves as

$$\mathbf{x}_{k+1} = A(T_S)\mathbf{x}_k + B(T_S)\mathbf{u}_k + \mathbf{w}_k, \qquad (4.4)$$

where

$$A(T_S) = e^{A_c T_S}$$
$$B(T_S) = \int_0^{T_S} e^{A_c \tau} B_c \mathrm{d}\tau$$

and $\mathbf{w}_k$ is a Gaussian white noise with zero mean and covariance

$$W(T_S) = \int_0^{T_S} e^{A_c \tau} W_c e^{A_c^\mathsf{T} \tau} \mathrm{d}\tau,$$

see also Demirel et al. (2011); Franklin et al. (1997).

As shown in Demirel et al. (2011); Levis et al. (1971), the continuous-time cost (4.2) can be calculated from the discrete-time system (4.4) as follows.

$$J(T_S) = \mathbf{x}_N^\mathsf{T} F \mathbf{x}_N + \sum_{k=0}^N c_k(T_S), \qquad (4.5)$$
$$c_k(T_S) = \mathbf{x}_k^\mathsf{T} Q(T_S)\mathbf{x}_k + 2\mathbf{x}_k^\mathsf{T} H(T_S)\mathbf{u}_k + \mathbf{u}_k^\mathsf{T} R(T_S)\mathbf{u}_k,$$

where $c_k(T_S)$ is the *cost per step*. We assume that $NT_S = T_F$ holds, i.e., T_F is an integer multiple of the sampling time. Moreover, $F = F_c$ and the matrices $Q(T_S)$, $H(T_S)$, and $R(T_S)$ are

$$Q(T_S) = \int_0^{T_S} e^{A_c^\mathsf{T} \tau} Q_c e^{A_c \tau} \mathrm{d}\tau,$$
$$H(T_S) = \int_0^{T_S} e^{A_c^\mathsf{T} \tau} Q_c \int_0^\tau e^{A_c s} \mathrm{d}s B_c \mathrm{d}\tau,$$
$$R(T_S) = T_S R_c + B_c^\mathsf{T} \int_0^{T_S} \int_0^\tau e^{A_c^\mathsf{T} s} \mathrm{d}s Q_c \int_0^\tau e^{A_c s} \mathrm{d}s \mathrm{d}\tau B_c.$$

In order to minimize the cost (4.2) of the continuous-time system (4.1), we now minimize the cost (4.5) of the corresponding discrete-time system (4.4).

Note that the cost (4.5) depends on the sampling time T_S. Interestingly, the minimal cost is not always increasing with the sampling time since controllability can get lost, as stated in the following theorem.

Theorem 4.1 (Kalman et al. (1963)). *Let the continuous-time system* (4.1) *be controllable. Then the discrete-time system* (4.4) *is controllable if:*

$$\text{Im}\big(\lambda_i(A_c) - \lambda_j(A_c)\big) \neq n\frac{2\pi}{T_s}, \quad n = \pm 1, \pm 2, \ldots$$

whenever

$$\text{Re}\big(\lambda_i(A_c) - \lambda_j(A_c)\big) = 0,$$

where $\lambda_i(A_c)$ *is the i-th eigenvalue of* A_c. *If the control is scalar, then the condition is necessary as well.*

Obviously, if the discrete-time system is not controllable, the minimal cost might be infinite.

4.3. Design of the Transport Layer

In the field of communication theory, it is well known that a reliable link can be realized by retransmitting lost packets until the corresponding acknowledgment arrives. Unfortunately, this approach has one drawback. Due to the acknowledgment mechanism the time between packet generation and reception of the corresponding acknowledgment is neither known in advance nor limited. Consequently, this approach is not directly applicable for real-time data. Nevertheless, we borrow this idea but limit the number of retransmissions. In general, sending an acknowledgement takes some time. Thus, it might be better to send data packets instead of acknowledgements. Hence, we introduce and analyze two transport layer protocols.

The first protocol uses acknowledgements and automatic retransmissions to increase the arrival probability. However, in contrast to classical transport layer protocols, the number of transmissions is limited to M to guarantee a deterministic behavior. Since this would be a slight modification of the error control mechanism of TCP, this protocol will be called *TCP-like*[1] within this chapter. Figure 4.3a shows an example of such a TCP-like protocol with up to $M = 3$ transmissions. Within this figure, the grey arrows indicate possible transmissions, which only take place when previous data packets or acknowledgment packets are lost. Moreover, note that the receiver waits as long as would be required to transmit all M data packets before it passes the received data to the next upper layer to guarantee a deterministic behavior.

In the second protocol, the arrival probability is increased by transmitting each packet M times. Since this is somewhat similar to UDP (no acknowledgements and

---

[1]Within the field of networked control systems, the term TCP-like protocol refers to a protocol with acknowledgements. However, TCP is a much more powerful protocol that offers not only error control but also flow control, segmentation, and connection management. Since the focus of this chapter is on the error correction within the transport layer, we use the term TCP-like protocol to refer to a transport layer protocol with acknowledgements and a limited number of retransmissions.

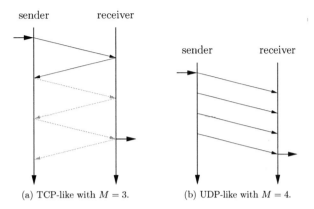

(a) TCP-like with $M = 3$.   (b) UDP-like with $M = 4$.

Figure 4.3.: TCP-like and UDP-like protocol.

retransmissions), this protocol will be called *UDP-like*[2] within this chapter. Fig 4.3b shows an example of such an UDP-like protocol with $M = 4$ transmissions. Again, the receiver waits as long as would be required to transmit all M data packets before it passes the received data to the next upper layer to guarantee a deterministic behavior.

With these two protocols, a more reliable communication system is build on top of an unreliable communication system. To distinguish between these communication systems, we use the terms *overlay* and *underlay* communication system. Notationally, we underline parameters of the underlay communication system and overline parameters of the overlay communication system. The following two lemmas give the overlay arrival probabilities as well as the expected number of packets for a TCP-like and an UDP-like connection. These values are contained in (Kögel, 2009, Theorem 7.5).

Lemma 4.2 (TCP-like). *Suppose, an overlay link is build by a TCP-like connection with up to M transmissions. The arrival probability of data packets is $\underline{p}_{\text{data}}$ and $\underline{p}_{\text{ACK}}$ for acknowledgement packets. Then,*

(i) the arrival probability of data packets is

$$\bar{p}_{\text{data}} = 1 - (1 - \underline{p}_{\text{data}})^M.$$

(ii) the arrival probability of acknowledgment packets is

$$\bar{p}_{\text{ACK}} = \frac{p_*}{1 - (1 - \underline{p}_{\text{data}})^M},$$

---

[2]Within the field of networked control systems, the term UDP-like protocol refers to a protocol without acknowledgements.

where $p_* = \sum_{i=0}^{M-1} (1 - \underline{p}_\text{data}\underline{p}_\text{ACK})^i \underline{p}_\text{data}\underline{p}_\text{ACK}$ *is the probability that an acknowledgment arrives at the sender, i.e., the probability that both, the data and the acknowledgment packet arrive.*

(iii) the probability that a data packet arrived at the receiver, given the fact that no acknowledgment arrived at the sender is

$$\bar{\epsilon} = \frac{1 - (1 - \underline{p}_\text{data})^M - p_*}{1 - p_*} = \frac{\bar{p}_\text{data}(1 - \bar{p}_\text{ACK})}{1 - \bar{p}_\text{data}\bar{p}_\text{ACK}}.$$

(iv) the expected number of sent data packets is

$$\mathrm{E}[N_\text{data}] = \sum_{j=0}^{M-1} (1 - \underline{p}_\text{data}\underline{p}_\text{ACK})^j.$$

(v) the expected number of sent acknowledgements is

$$\mathrm{E}[N_\text{ACK}] = \underline{p}_\text{data} \sum_{j=0}^{M-1} (1 - \underline{p}_\text{data}\underline{p}_\text{ACK})^j.$$

Lemma 4.3 (UDP-like). *Suppose, an overlay link is build by an UDP-like connection with M transmissions. The arrival probability of data packets is $\underline{p}_\text{data}$. Then, the arrival probability of data packets is*

$$\bar{p}_\text{data} = 1 - (1 - \underline{p}_\text{data})^M.$$

Retransmitting packets or sending them multiple times has some consequences for the choice of the sampling time. Since we do not want to increase the packet rate, increasing the number of transmissions per sampling interval requires to increase the sampling time. Moreover, note that for a networked control system, two connections must be designed: the connection between sensor and controller and the connection between controller and actuator. Thus, we consider four different configurations of the transport layer. Figure 4.4 shows the timing of these configurations.

In the *TCP-TCP configuration* there is a TCP-like connection between sensor and controller and also a TCP-like connection between controller and actuator. Figure 4.4a shows an example of such a configuration with $M_\text{sc} = 2$ and $M_\text{ca} = 3$. However, within the considered setup, there is no benefit of acknowledging measurement packets. Thus, by using an UDP-like connection between sensor and controller it is possible to send more measurement packets per time. Figure 4.4b shows an example of such an *UDP-TCP configuration* with $M_\text{sc} = 3$ and $M_\text{ca} = 3$. When comparing Figure 4.4a with Figure 4.4b, we see that the sensor sends one more measurement packet, although the sampling time is equal. Thus, the probability that a measurement arrives at the controller is larger. Similarly, it is also possible to use an UDP-like connection between

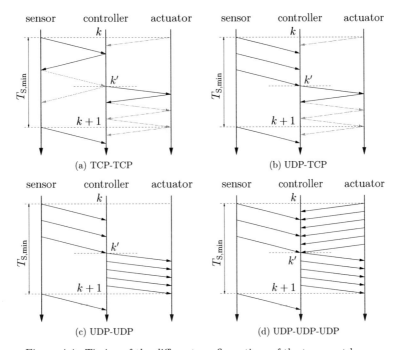

Figure 4.4.: Timing of the different configurations of the transport layer.

controller and actuator. In doing so, it is possible to send more control packets per time. This results in a higher probability that a control packet arrives at the actuator but the controller does not know whether or not a control packet arrived. Since this knowledge is important for a good estimator and controller design, see, e.g. Schenato et al. (2007), it is not clear whether this tradeoff is worth. Figure 4.4c shows an example of such an *UDP-UDP configuration* with $M_{sc} = 3$ and $M_{ca} = 5$. Finally, Figure 4.4d depicts an example of the *UDP-UDP-UDP configuration*, where the time between k and k' is used to create a dedicated UDP-like connection between the actuator and controller to send the information whether or not the control packet arrived.

Note that in the UDP-UDP-UDP case, the acknowledgements of control packets are always sent and contain the information whether or not a control packet arrived, i.e., positive and negative acknowledgements (ACKs and NAKs) are sent. In contrast, in the TCP-TCP and UDP-TCP case, the acknowledgements are only sent when the control packet arrived, i.e., only positive acknowledgments (ACKs) are sent.

Table 4.1 summarizes the minimal sampling time, the arrival probabilities, and the expected number of packets per sampling interval for the four different configurations of the transport layer. Thereby, M is the maximal number of transmissions and $E[N]$ the expected number of transmissions.

When looking at Table 4.1, we see that the arrival probability of measurement packets is identical in all four configurations, i.e., it depends only on the maximal number of transmissions and not on the configuration. The same holds for the arrival probability of control packets. Nevertheless, the minimal sampling time depends on the choice of the configuration. Given the same number of transmissions per sampling interval, the minimal sampling time of TCP-TCP is larger than the one of UDP-TCP, which is larger than the one of UDP-UDP and UDP-UDP-UDP. From another point of view this means that for a given sampling time more measurement and control packets can be transmitted per sampling interval and thus a higher arrival probability can be achieved when an UDP-like connection is used. Thus, the cost achieved with the TCP-TCP configuration is larger than the cost achieved with the UDP-TCP configuration, which is larger than the cost achieved with the UDP-UDP-UDP configuration, i.e.,

$$J_{\text{TCP-TCP}} \geq J_{\text{UDP-TCP}} \geq J_{\text{UDP-UDP-UDP}}.$$

Due to the lack of acknowledgements of control packets in the UDP-UDP configuration, it is not clear how the performance of the UDP-UDP configuration compares with the performance of the UDP-TCP configuration.

On the other hand, when comparing a TCP-like and an UDP-like connection we have to take into account that for a TCP-like connection the transmission of packets is stopped when an acknowledgement arrives. Consequently, when a TCP-like connection is used, the expected number of transmitted packets is smaller than the maximal number of transmissions. Thus, for a given sampling time and the assumption that the delay of an acknowledgement packet and the delay of a data packet are equal, the expected number of transmitted packets for the TCP-TCP configuration is smaller than the one of the UDP-TCP configuration, which is smaller than the one of the UDP-UDP configuration, which is smaller than the one of the UDP-UDP-UDP configuration, i.e.,

$$E[N_{\text{TCP-TCP}}] \leq E[N_{\text{UDP-TCP}}] \leq N_{\text{UDP-UDP}} \leq N_{\text{UDP-UDP-UDP}}.$$

Finally, note that for $M_{\text{sc}} = M_{\text{ca}} = M_{\text{ac}}$, the packet rate of the UDP-UDP-UDP configuration is identical to the packet rate of the standard configuration.

4.4. Controller Design

Table 4.1 shows that a more reliable overlay link can be built by retransmitting lost packets or sending more than one data packet during each sampling interval. For simplicity of notation, we thus consider only the overlay communications system in the following. Therefore, we use the mutually independent and identically distributed

Table 4.1.: The minimal sampling time, the arrival probabilities, and the expected number of sent packets of the different configurations of the transport layer.

	TCP-TCP	UDP-TCP	UDP-UDP	UDP-UDP-UDP
$T_{S,\min}$	$\max\{\underline{d}_{sc} + (M_{sc}-1)$ $(\underline{d}_{sc}+\underline{d}_{cs}), \underline{d}_{ac}\} + \underline{d}_{ca}$ $+(M_{ca}-1)(\underline{d}_{ac}+\underline{d}_{ca})$	$\max\{M_{sc}\underline{d}_{sc}, \underline{d}_{ac}\} + \underline{d}_{ca}$ $+(M_{ca}-1)(\underline{d}_{ac}+\underline{d}_{ca})$	$\underline{d}_{sc}M_{sc} + \underline{d}_{ca}M_{ca}$	$\max\{\underline{d}_{sc}M_{sc}, \underline{d}_{ac}M_{ac}\}$ $+\underline{d}_{ca}M_{ca}$
$\bar{p}_{sc}$	$1-(1-\underline{p}_{sc})^{M_{sc}}$	$1-(1-\underline{p}_{sc})^{M_{sc}}$	$1-(1-\underline{p}_{sc})^{M_{sc}}$	$1-(1-\underline{p}_{sc})^{M_{sc}}$
$\bar{p}_{ca}$	$1-(1-\underline{p}_{ca})^{M_{ca}}$	$1-(1-\underline{p}_{ca})^{M_{ca}}$	$1-(1-\underline{p}_{sc})^{M_{sc}}$	$1-(1-\underline{p}_{ca})^{M_{ca}}$
$\bar{p}_{ac}$	$\dfrac{\sum_{i=0}^{M_{ca}-1}(1-\underline{p}_{ca}\underline{p}_{ac})^i\underline{p}_{ca}\underline{p}_{ac}}{1-(1-\underline{p}_{ca})^{M_{ca}}}$	$\dfrac{\sum_{i=0}^{M_{ca}-1}(1-\underline{p}_{ca}\underline{p}_{ac})^i\underline{p}_{ca}\underline{p}_{ac}}{1-(1-\underline{p}_{ca})^{M_{ca}}}$	—	$1-(1-\underline{p}_{ac})^{M_{ac}}$
$\bar{\epsilon}$	$\dfrac{\bar{p}_{ca}(1-\bar{p}_{ac})}{1-\bar{p}_{ca}\bar{p}_{ac}}$	$\dfrac{\bar{p}_{ca}(1-\bar{p}_{ac})}{1-\bar{p}_{ca}\bar{p}_{ac}}$	$\bar{p}_{ca}$	$\bar{p}_{ca}$
$E[N_{sc}]$	$\sum_{j=0}^{M_{sc}-1}(1-\underline{p}_{sc}\underline{p}_{cs})^j$	M_{sc}	M_{sc}	M_{sc}
$E[N_{cs}]$	$\underline{p}_{sc}\sum_{j=0}^{M_{sc}-1}(1-\underline{p}_{sc}\underline{p}_{cs})^j$	0	0	0
$E[N_{ca}]$	$\sum_{j=0}^{M_{ca}-1}(1-\underline{p}_{ca}\underline{p}_{ac})^j$	$\sum_{j=0}^{M_{ca}-1}(1-\underline{p}_{ca}\underline{p}_{ac})^j$	M_{ca}	M_{ca}
$E[N_{ac}]$	$\underline{p}_{ac}\sum_{j=0}^{M_{ca}-1}(1-\underline{p}_{ca}\underline{p}_{ac})^j$	$\underline{p}_{ac}\sum_{j=0}^{M_{ca}-1}(1-\underline{p}_{ca}\underline{p}_{ac})^j$	0	M_{ac}

Bernoulli random processes $\{\gamma_k\} \in \{0,1\}$, $\{\beta_k\} \in \{0,1\}$, and $\{\text{ACK}_k\} \in \{0,1\}$ with $\Pr(\gamma_k = 1) = \bar{p}_{\text{sc}}$, $\Pr(\beta_k = 1) = \bar{p}_{\text{ca}}$, and $\Pr(\text{ACK}_k = 1) = \bar{p}_{\text{ac}}$ to describe the arrival of measurement, control, and acknowledgment packets. In doing so, it is possible to use the well established theory of optimal control over lossy communication links.

As in Garone et al. (2008); Imer et al. (2006); Kögel (2009); Kögel, Blind, and Allgöwer (2010); Schenato et al. (2007), we assume that the input is kept zero, i.e., $\mathbf{u}_k = 0$ whenever the actuator does not receive a control packet. Obviously, there exist other strategies, e.g., reusing the previous input, see Schenato (2009) for a more detailed discussion. To distinguish between the control input applied to the plant and the control input calculated by the controller, we use the superscript c to indicate that the control input $\mathbf{u}^c$ is calculated by the controller, whereas $\mathbf{u}$ remains the control input applied to the plant. Due to the packet loss, we have $\mathbf{u}_k = \beta_k \mathbf{u}_k^c$. Moreover, from the point of view of the controller, the state of the plant evolves as

$$\mathbf{x}_{k+1} = A(T_{\text{S}})\mathbf{x}_k + \beta_k B(T_{\text{S}})\mathbf{u}_k^c + \mathbf{w}_k, \tag{4.6}$$

and the cost becomes

$$J(T_{\text{S}}) = \mathbf{x}_N^{\mathsf{T}} F \mathbf{x}_N + \sum_{k=0}^{N} c_k(T_{\text{S}}), \tag{4.7}$$
$$c_k(T_{\text{S}}) = \mathbf{x}_k^{\mathsf{T}} Q(T_{\text{S}})\mathbf{x}_k + 2\beta_k \mathbf{x}_k^{\mathsf{T}} H(T_{\text{S}})\mathbf{u}_k^c + \beta_k \mathbf{u}_k^{c\mathsf{T}} R(T_{\text{S}})\mathbf{u}_k^c.$$

From Figure 4.4, we see that the next control input $\mathbf{u}_{k+1}^c$ is calculated at time k' and thus must be calculated based on the knowledge of the state $\mathbf{x}_k$ (and all previous states) and the current acknowledgment ACK_k (and all older acknowledgments). Consequently, a predictive control scheme must be used. Therefore, the state is predicted as in Garone et al. (2008); Kögel (2009); Kögel, Blind, and Allgöwer (2010).

Measurement update:

$$\hat{\mathbf{x}}_{k|k} = \gamma_k \mathbf{x}_k + (1 - \gamma_k)\hat{\mathbf{x}}_{k|k-1}, \quad \hat{\mathbf{x}}_{0|-1} = \bar{\mathbf{x}}_0.$$

Time update:

$$\hat{\mathbf{x}}_{k+1|k} = A\hat{\mathbf{x}}_{k|k} + \mathrm{E}[\beta_k|\text{ACK}_k]B\mathbf{u}_k^c,$$

where

$$\mathrm{E}[\beta_k|\text{ACK}_k] = \begin{cases} \beta_k & \text{if } \text{ACK}_k = 1 \\ \bar{\epsilon} & \text{if } \text{ACK}_k = 0. \end{cases}$$

Combining the measurement and time update, we get

$$\hat{\mathbf{x}}_{k+1|k} = \gamma_k A\mathbf{x}_k + (1 - \gamma_k)A\hat{\mathbf{x}}_{k|k-1} + \mathrm{E}[\beta_k|\text{ACK}_k]B\mathbf{u}_k^c.$$

Now, the optimal control law is given in the following theorem.

Theorem 4.4 (Kögel (2009); Kögel, Blind, and Allgöwer (2010)). *Consider the system (4.6) with cost (4.7) and arrival probabilities $\overline{p}_{\mathrm{sc}}$, $\overline{p}_{\mathrm{ac}}$, and $\overline{p}_{\mathrm{ca}}$. Then the optimal control law is linear and given by*

$$\mathbf{u}_k^{\mathrm{c}} = -K_k \hat{\mathbf{x}}_{k|k-1},$$
$$K_k = (R + B^{\mathsf{T}}(S_{k+1} + \phi P_{k+1})B)^{-1}(B^{\mathsf{T}}S_{k+1}A + H^{\mathsf{T}}),$$
$$P_k = (1 - \overline{p}_{\mathrm{sc}})A^{\mathsf{T}}P_{k+1}A + \overline{p}_{\mathrm{ca}}(A^{\mathsf{T}}S_{k+1}B + H)K_k,$$
$$S_k = A^{\mathsf{T}}S_{k+1}A + Q - \overline{p}_{\mathrm{ca}}(A^{\mathsf{T}}S_{k+1}B + H)K_k,$$
$$S_N = F,$$
$$P_N = 0,$$

with

$$\phi = \begin{cases} 1 - \overline{p}_{\mathrm{ca}} & \textit{UDP-UDP,} \\ \dfrac{(1-\overline{p}_{\mathrm{ac}})(1-\overline{p}_{\mathrm{ca}})}{1-\overline{p}_{\mathrm{ac}}\overline{p}_{\mathrm{ca}}} & \textit{TCP-TCP and TCP-UDP,} \\ (1 - \overline{p}_{\mathrm{ac}})(1 - \overline{p}_{\mathrm{ca}}) & \textit{UDP-UDP-UDP.} \end{cases}$$

Furthermore, the expected cost is

$$J_N = \bar{\mathbf{x}}_0^{\mathsf{T}} S_0 \bar{\mathbf{x}}_0 + \mathrm{Tr}\big((S_0 + P_0)X_0\big) + \sum_{i=1}^{N} \mathrm{Tr}\big((S_i + P_i)W\big).$$

Moreover, if this iteration converges, then there exists an infinite horizon controller with a finite cost per step

$$c_{\infty} = \lim_{N \to \infty} \frac{1}{N} J_N = \mathrm{Tr}\left((S_{\infty} + P_{\infty})W\right).$$

Note that the previous equation gives the cost per step, but the step size is given by the sampling time. In order to compare different configurations fairly we define the relative cost per step $\tilde{c}_{\infty}$ as the cost per step divided by the sampling time, i.e.,

$$\tilde{c}_{\infty} := c_{\infty}/T_{\mathrm{S}}.$$

Unfortunately, it is not clear whether the iteration of Theorem 4.4 converges for given arrival probabilities. When the arrival probability is one and all packets arrive, the considered problem simplifies to the standard LQR problem, which converges as long as the discrete-time system is stabilizable and the weights are properly chosen. On the other extreme, when the arrival probability is zero and all packets are lost, we have $P_k = A^{\mathsf{T}}P_{k+1}A$ and $S_k = A^{\mathsf{T}}S_{k+1}A + Q$. In this case, the iteration will not converge for unstable systems.

The most pragmatic way to find the stabilizable region, i.e., the arrival probabilities for which there exists a stabilizing controller, is to simply start the iteration of Theorem 4.4 and check whether it converges. In doing so, convergence is easily detected.

However, the decision that the iteration does not converge is not that simple. It can only be conjectured from the fact that there is no convergence after a finite number of iterations or when the resulting cost is extremely large. Fortunately, an easier to check necessary condition is given in Kögel (2009).

Theorem 4.5 (Kögel (2009)). *Assume (A, B) stabilizable, $(A, Q^{1/2})$ observable, $R > 0$. A necessary condition for the existence of a stable optimal predictive controller is*

$$1 - (1 - \bar{p}_{\mathrm{sc}})\lambda_{\max}^2 > 0$$

$$\lambda_{\max}^2 \left(1 - \frac{\bar{p}_{\mathrm{ca}}}{1 - \phi + \frac{\phi \bar{p}_{\mathrm{sc}} \lambda_{\max}^2}{1 - (1 - \bar{p}_{\mathrm{sc}})\lambda_{\max}^2}} \right) < 1,$$

where $\lambda_{\max}$ is the largest magnitude of the eigenvalues of A.

If B is invertible, then this condition is necessary and sufficient.

4.5. Examples

In this section, we look at three examples to discuss whether it is worth to increase the sampling time to allow retransmitting lost packets. In the first example, we show that the stabilizable region can be increased by a proper choice of the number of transmissions. Within the second example, we look how the cost per step depends on the arrival probability, the sampling time, and the configuration of the transport layer. Finally, the last example considers a slightly more complex system, where the optimal configuration of the communication system and the optimal choice of the sampling time is not that simple anymore.

Example 1

Within this example, we check how the stabilizable region depends on the number of transmissions for the UDP-UDP setup. Therefore, we use a simple scalar system $A_c = 1$, $B_c = 1$ and delays $\underline{d}_{\mathrm{sc}} = \underline{d}_{\mathrm{ca}} = 0.1$. This system is sampled as fast as possible, i.e., with $T_{\mathrm{S,min}} = \underline{d}_{\mathrm{sc}} M_{\mathrm{sc}} + \underline{d}_{\mathrm{ca}} M_{\mathrm{ca}}$ as given in Table 4.1.

Figure 4.5a shows the stability region for $M_{\mathrm{ca}} = M_{\mathrm{sc}} = M$. Interestingly, for $\underline{p}_{\mathrm{sc}} = \underline{p}_{\mathrm{ac}}$ the number of transmissions does not affect the stabilizable region. However, for $\underline{p}_{\mathrm{sc}} < \underline{p}_{\mathrm{ca}}$, it is possible to increase the stabilizable region by increasing the number of transmissions. On the other hand, for $\underline{p}_{\mathrm{sc}} > \underline{p}_{\mathrm{ca}}$, the stabilizable region shrinks when the number of transmissions per sampling interval is increased. Obviously, restricting the number of measurement packets to be equal to the number of control packets, i.e., $M_{\mathrm{sc}} = M_{\mathrm{ca}}$ is not necessarily optimal. Thus Figure 4.5b shows the stabilizable region for $M_{\mathrm{ca}} = 2M_{\mathrm{sc}}$, i.e., during each sampling interval twice as many control packets as measurement packets are sent. Now, the stabilizable region can be increased for $\underline{p}_{\mathrm{sc}} = \underline{p}_{\mathrm{ca}}$ by increasing the number of transmissions per sampling interval. However, for

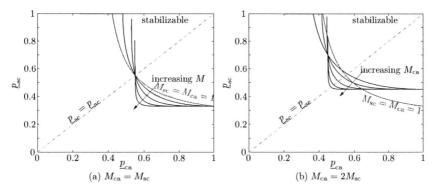

Figure 4.5.: Stabilizable region for $M_{ca} = M_{sc}$, $M_{ca} = 2M_{sc}$.

other arrival probabilities, the stabilizable region shrinks with an increasing number of transmissions per sampling interval. Finally, Figure 4.6 shows the Pareto frontier of the arrival probabilities for which a stabilizing controller can be found with $M_{sc} \leq 16$ and $M_{ca} \leq 16$. Here, we clearly see that by a proper choice of the number of transmissions, the stabilizable region can be increased significantly.

From this example, we conclude that by a proper design of the transport layer, it is possible to stabilize a system, which can not be stabilized when only one measurement and one control packet is sent during each sampling interval. Remember that our approach does not require to increase the number of packets per time.

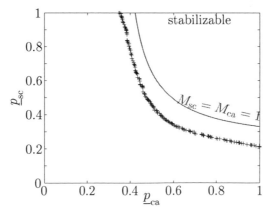

Figure 4.6.: Stabilizable region for $M_{ca} \leq 16$ and $M_{sc} \leq 16$.

Example 2

In this example, we raise the question whether it is possible to improve the performance by increasing the sampling time to allow more transmissions per sampling interval. Again, we use a simple scalar system with $A_c = 1$, $B_c = 1$, $W_c = 1$, $Q_c = 1$, $R_c = 1$, $\underline{d}_{sc} = \underline{d}_{cs} = \underline{d}_{ca} = \underline{d}_{ac} = 0.01$, and assume that all underlay arrival probabilities are equal, i.e., $\underline{p}_{sc} = \underline{p}_{cs} = \underline{p}_{ca} = \underline{p}_{ac} = p$. Finally, we assume that the maximal number of transmissions are equal, i.e., $M_{sc} = M_{ca} = M$ but keep in mind that this might not be optimal. Moreover, the system is sampled as fast as possible, i.e., with $T_{S,min}$ as given in Table 4.1. Thus, the sampling time follows directly from the maximal number of transmissions M.

Figure 4.7 shows the relative cost per step $\tilde{c}_{\infty}$ over the sampling time T_S for a fixed arrival probability $\underline{p}$ of 0.3 and 0.7, respectively. Moreover, Figure 4.8 shows how the relative cost per step depends on the arrival probability and the sampling time of the four considered configurations.

From these figures, we see that it can be worth to increase the sampling time to allow more transmissions per sampling interval. From Figure 4.7a, we see that the performance can be improved significantly by a proper choice of the sampling time when the reliability of the communication system is weak. On the other hand, when the communication system is already relatively reliable, as in Figure 4.7b, the benefit of a proper choice of the sampling time becomes less significant.

From Figure 4.7, we also see that the relative cost per step for the TCP-TCP configuration is larger than the one of the UDP-TCP configuration, which is larger than the one of the UDP-UDP-UDP configuration, as already stated. Moreover, we see that the relative cost per step achieved with the UDP-UDP configuration is only slightly larger than the one of the UDP-UDP-UDP configuration.

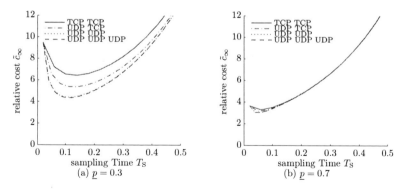

Figure 4.7.: The relative cost per step over the sampling time T_S for Example 2.

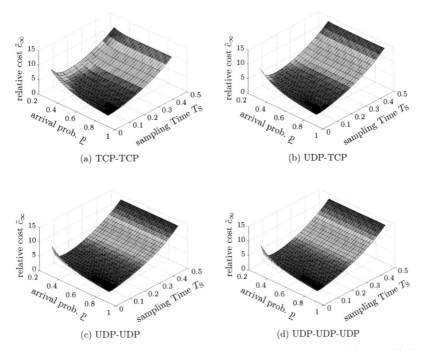

(a) TCP-TCP

(b) UDP-TCP

(c) UDP-UDP

(d) UDP-UDP-UDP

Figure 4.8.: The relative cost per step over the underlay arrival probability $\underline{p}$ and the sampling time T_S for Example 2.

Example 3

Finally, we look at a slightly more complex example with

$$A_c = \begin{bmatrix} 0 & \omega \\ -\omega & -2\xi\omega \end{bmatrix}, \quad B_c = \begin{bmatrix} 0 \\ 1 \end{bmatrix}$$

and $Q_c = I$, $R_c = 1$, and $W_c = I$.

Note that the eigenvalues of A_c are

$$\lambda_{1,2} = -\xi\omega \pm j\omega\sqrt{1-\xi^2}.$$

By choosing $\omega = \frac{1}{2}$ and $\xi = -\frac{1}{2}$, the eigenvalues become $\lambda_{1,2} = \frac{1}{4} \pm \frac{\sqrt{3}}{4}j$, and the critical sampling times as given in Theorem 4.1 are $\frac{4\pi}{\sqrt{3}}n \approx 7.255\,n$, $n = 1, 2, \ldots$. Moreover, the delays are assumed to be $\underline{d}_{sc} = \underline{d}_{cs} = \underline{d}_{ca} = \underline{d}_{ac} = 0.1$ and we assume that all underlay arrival probabilities are equal, i.e., $\underline{p}_{sc} = \underline{p}_{cs} = \underline{p}_{ca} = \underline{p}_{ac} = \underline{p}$. Again, we assume that the maximal number of transmissions are equal, i.e., $M_{sc} = M_{ca} = M$ and that the system is sampled as fast as possible, i.e., with $T_{S,min}$ as given in Table 4.1. Again, the sampling time follows directly from the maximal number of transmissions M.

Figure 4.9 shows how the relative cost per step depends on the arrival probability $\underline{p}$ and the sampling time T_S. While calculating the relative cost per step, we stopped the iteration when $\|P + S\|$ of Theorem 4.4 exceeded 10^{15}. Thus, the plateaus on the upper left sides are the area for which we could not find a stabilizing controller. Not surprisingly, the smallest arrival probability, for which we found a stabilizing controller depends on the sampling time T_S and the configuration of the transport layer. Moreover, we also see that the relative cost per step increases when the sampling time T_S is close to the critical sampling time.

Figure 4.10 shows the relative cost per step over the sampling time T_S for an arrival probability of 0.2 and 0.3. When looking at Figure 4.10a ($\underline{p} = 0.2$), we see that no stabilizing controller could be found for the TCP-TCP, UDP-TCP, or UDP-UDP configuration. However, with the UDP-UDP-UDP configuration and a proper choice of the sampling time, a stabilizing controller could be found. Remember that the difference between the UDP-UDP and the UDP-UDP-UDP configuration is only the availability of acknowledgments of the control packets. Thus, this example also demonstrates how important these acknowledgments can be.

When looking at Figure 4.10b ($\underline{p} = 0.3$), we see a much larger difference between the different configurations. For the TCP-TCP configuration, a stabilizing controller exists only for $M = 1$, which is indeed identical to the standard fast sampling. If the sampling time is increased to allow retransmitting lost packets, no stabilizing controller can be found. Since for $M = 1$, there is no difference between the TCP-TCP and the UDP-TCP configuration, there also exists a stabilizing controller for the UDP-TCP configuration with the same relative cost per step. However, with the UDP-TCP configuration, it is also possible to stabilize the system with a longer sampling time; more precisely, for $M = \{11, 12, 13, 14\}$. Nevertheless, the minimal relative cost per

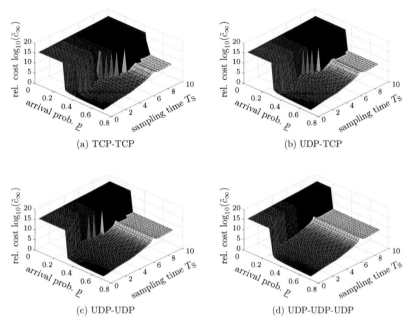

Figure 4.9.: The relative cost per step over the underlay arrival probability $\underline{p}$ and the sampling time T_S for Example 3.

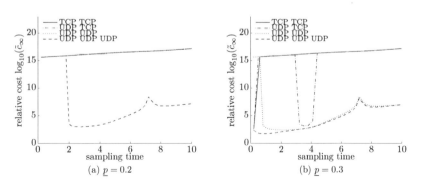

Figure 4.10.: The relative cost per step over the sampling time T_S for Example 3.

step is achieved for $M = 1$. When looking at the UDP-UDP configuration, we see that there exists no stabilizing controller for $M = 1$. For $M = 1$, the difference between the TCP-TCP, UDP-TCP, and UDP-UDP configuration is the lack of an acknowledgement in the UDP-UDP configuration. Again, this observation indicates the crucial role of acknowledgements. However, for larger values of M, there exists a stabilizing controller for the UDP-UDP configuration with a slightly smaller relative cost per step than the one achieved with the standard fast sampling. Finally, we see that the UDP-UDP-UDP configuration outperforms all other schemes. With the proper choice of the sampling time, a better performance than for the standard fast sampling scheme can be achieved.

4.6. Summary

We analyzed the effect of retransmitting lost packets on the control performance. Obviously, by retransmitting lost packets, the communication system becomes more reliable. However, to allow retransmitting lost packets either the packet rate or the sampling time must be increased. We argued that increasing the packet rate might be problematic and chose to increase the sampling time instead. Now, increasing the sampling time increases the reliability of the communication system but decreases the control performance. Due to this tradeoff, the choice of the sampling time becomes an interesting problem. For four different setups (TCP-TCP, UDP-TCP, UDP-UDP, and UDP-UDP-UDP), we derived the minimal sampling time as well as the overlay arrival probabilities depending on the maximal number of transmissions per sampling interval. In doing so, the problem was reformulated as the optimal control problem over lossy links. Since this problem is already well studied, we could build on these results but still had the freedom to choose the sampling time. With the help of three examples, we finally showed that it is possible to increase the stabilizable region and improve the control performance by a proper choice of the sampling time.

Chapter 5.

Optimization within the Network Layer: Routing and Controller Placement

In communication systems, the main task of the network layer is the routing, i.e., finding an appropriate route from a source to a destination node. In the Internet, the route is often chosen such that the delay or the number of hops is minimized, see, e.g., Tanenbaum (2003). Obviously, for networked control systems, this is not necessarily the best approach. Moreover, in networked control systems, not only the route from the sensor to the controller and the route from the controller to the actuator must be chosen, but also the location of the controller might be a degree of freedom. Thus, when considering huge communication systems, with many nodes and links, one interesting question is the optimal position of the controller within the communication system and the optimal routes between sensor, controller, and actuator. Hence, we consider the joint design of the controller, its placement within the communication system, and the routing through the communication system within this chapter. This chapter is based on Blind and Allgöwer (2013c).

5.1. Problem Setup

The considered problem is relatively similar to the one of the previous chapter. As in the previous chapter, we assume that the sampling time and arrival probabilities depend on the design of the communication system. In contrast to the previous chapter, where the end-to-end delays and arrival probabilities follow from the design of the transport layer, they follow from the design of the network layer within this chapter. Hence, the end-to-end delays and arrival probabilities are determined by the choice of the routes between sensor, controller, and actuator and we search for the optimal position of the controller and the optimal routes between sensor, controller, and actuator.

A somewhat similar problem is studied in Robinson and Kumar (2008) and Quevedo et al. (2012). For full state measurements, it is shown in Robinson and Kumar (2008) that the best performance is achieved when the controller is located at the actuator.

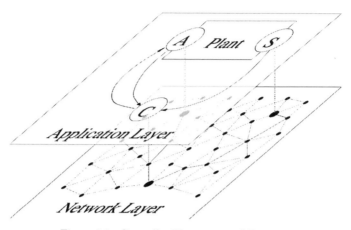

Figure 5.1.: Controller Placement and Routing.

In Quevedo et al. (2012), the controller is dynamically allocated and thus its location depends on the losses along the route. Thereby, all nodes compute local state estimates. However, both works assume that the delays introduced by the network can be neglected.

A related problem is studied in Carabelli et al. (2012), where several operators and their connection, which are described by an *overlay network*, must be placed and routed within an *underlay network* such that the bandwidth-delay product is minimized. We use this idea to study the joint design of controller and routing for networked control systems. In doing so, the overlay network is given by the control structure and the underlay network is given by the underlying communication system. Figure 5.1 depicts this idea. In contrast to Carabelli et al. (2012), the cost function is no longer the delay-bandwidth product but follows from the control performance. As in the previous chapter, we use an underline for variables of the underlay network and an overline for variables of the overlay network.

5.1.1. Control System

As in the previous chapter, we consider the control of a continuous-time system

$$\dot{\mathbf{x}}(t) = A_c \mathbf{x}(t) + B_c \mathbf{u}(t) + \mathbf{w}(t), \tag{5.1a}$$
$$\mathbf{y}(t) = C_c \mathbf{x}(t) + \mathbf{v}(t), \tag{5.1b}$$

where $\mathbf{x} \in \mathbb{R}^{n_x}$ is the state of the system, $\mathbf{u} \in \mathbb{R}^{n_u}$ the control input, $\mathbf{y} \in \mathbb{R}^{n_y}$ the measurement, $\mathbf{w} \in \mathbb{R}^{n_x}$ the process noise, and $\mathbf{v} \in \mathbb{R}^{n_y}$ the measurement noise. The process noise $\mathbf{w}$ and measurement noise $\mathbf{v}$ are assumed to be mutually independent and Gaussian with zero mean and covariance $W_c \in \mathbb{R}^{n_x \times n_x}$ and $V_c \in \mathbb{R}^{n_y \times n_y}$, respectively.

The goal is to design a controller that minimizes the cost

$$J = \mathbf{x}^\mathsf{T}(T_\mathrm{F})F_c\mathbf{x}(T_\mathrm{F}) + \int_0^{T_\mathrm{F}} \mathbf{x}^\mathsf{T}(t)Q_c\mathbf{x}(t) + \mathbf{u}^\mathsf{T}(t)R_c\mathbf{u}(t)\mathrm{d}t. \tag{5.2}$$

5.1.2. Communication System

The model of the communication system is based on the one of Carabelli et al. (2012). It is modeled as a graph $\mathcal{G} = (\mathcal{N}, \mathcal{E})$, where $\mathcal{N}$ represents the node set and $\mathcal{E}$ the link (or edge) set. When sending a packet over a link, it will be delayed or even lost. Thus, we use $\underline{d}_i$ to denote the delay of link i and collect all delays in the vector $\underline{\mathbf{d}} := \left[\underline{d}_1, \ldots, \underline{d}_{|\mathcal{E}|}\right]^\mathsf{T}$. Moreover, the arrival probability of link i is denoted by $\underline{p}_i$. All arrival probabilities are collected in the vector $\underline{\mathbf{p}} := \left[\underline{p}_1, \ldots, \underline{p}_{|\mathcal{E}|}\right]^\mathsf{T}$.

The end-to-end delay of a route between start node s and end node e is just the sum of the delays of the links along the route, i.e.,

$$\overline{d}_{se} = \sum_{i \in \text{traversed links}} \underline{d}_i. \tag{5.3}$$

Similarly, the end-to-end arrival probability of a route is the product of the arrival probabilities of the links along the route, i.e.,

$$\overline{p}_{se} = \prod_{i \in \text{traversed links}} \underline{p}_i. \tag{5.4}$$

The topology of the communication network is described by the incidence matrix $E \in \{0, \pm 1\}^{\mathcal{N} \times \mathcal{E}}$. Each column of this incidence matrix represents a link. The entries of the columns are 1 when the corresponding node is the source, -1 when the corresponding node is the destination, and 0 otherwise.

To indicate the position of an overlay node j within the underlay network, we use the position indication vector $\mathbf{z}_{\mathrm{pos},j} \in \{0, 1\}^{|\mathcal{N}|}$. The vector $\mathbf{z}_{\mathrm{pos},j}$ contains only one non-zero entry 1, which indicates the position of overlay node j within the underlay network, i.e.,

$$[\mathbf{z}_{\mathrm{pos},j}]_i = \begin{cases} 1 & \text{overlay node } j \text{ is placed at underlay node } i, \\ 0 & \text{else.} \end{cases}$$

Moreover, we use the route indication vector $\mathbf{z}_{se} \in \{0, 1\}^{|\mathcal{E}|}$ to describe the route between overlay node s and e. The elements of $\mathbf{z}_{se}$ are one when the corresponding link is part of the route and zero otherwise, i.e.,

$$[\mathbf{z}_{se}]_l = \begin{cases} 1 & \text{link } l \text{ is part of the route,} \\ 0 & \text{else.} \end{cases}$$

Using this notation, the route and position indication vectors are related as follows

$$E\mathbf{z}_{se} = \mathbf{z}_{\mathrm{pos},s} - \mathbf{z}_{\mathrm{pos},e}.$$

Moreover, the end-to-end delay is now just the scalar product of the delay vector $\underline{\mathbf{d}}$ and the route indication vector $\mathbf{z}_{se}$, i.e., (5.3) can be written as

$$\bar{d}_{se} = \underline{\mathbf{d}}^\mathsf{T} \mathbf{z}_{se}.$$

Unfortunately, we can not rewrite (5.4) similarly due to the product. However, by using the logarithm, (5.4) can be written as

$$\log(\bar{p}_{se}) = \sum_{i \in \text{traversed links}} \log(\underline{p}_i).$$

By using $\underline{\tilde{\mathbf{p}}} := \left[\log(\underline{p}_1), \ldots, \log(\underline{p}_{|\mathcal{E}|}) \right]^\mathsf{T}$ to collect the logarithm of the arrival probabilities, we get for the end-to-end arrival probability

$$\log(\bar{p}_{se}) = \underline{\tilde{\mathbf{p}}}^\mathsf{T} \mathbf{z}_{se}.$$

5.2. Joint Design

5.2.1. Discretization

As discussed in the previous chapter, we use a discrete-time controller since the loop is closed by a packet based communication system. Again, system (5.1) is sampled with a constant sampling time T_S and it is assumed that the sensor and actuator are synchronized. To keep this chapter self containing, we shortly repeat the main steps of this discretization but refer to the previous chapter for the details. Sampling system (5.1) with the sampling time T_S gives the following discrete-time system.

$$\mathbf{x}_{k+1} = A(T_\mathrm{S})\mathbf{x}_k + B(T_\mathrm{S})\mathbf{u}_k + \mathbf{w}_k, \tag{5.5a}$$
$$\mathbf{y}_{k+1} = C\mathbf{x}_k + \mathbf{v}_k, \tag{5.5b}$$

and corresponding cost function

$$J(T_\mathrm{S}) = \mathbf{x}_N^\mathsf{T} F \mathbf{x}_N + \sum_{k=0}^{N} c_k(T_\mathrm{S}), \tag{5.6}$$
$$c_k(T_\mathrm{S}) = \mathbf{x}_k^\mathsf{T} Q(T_\mathrm{S})\mathbf{x}_k + 2\mathbf{x}_k^\mathsf{T} H(T_\mathrm{S})\mathbf{u}_k + \mathbf{u}_k^\mathsf{T} R(T_\mathrm{S})\mathbf{u}_k,$$

where $c_k(T_\mathrm{S})$ is the cost per step.

Thereby, $C = C_\mathrm{c}$, $F = F_\mathrm{c}$, and $V = V_\mathrm{c}$. The matrices A, B, W, Q, H, R are

$$A(T_\mathrm{S}) = e^{A_\mathrm{c} T_\mathrm{S}},$$

$$B(T_{\mathrm{s}}) = \int_0^{T_{\mathrm{s}}} e^{A_{\mathrm{c}}\tau} B_{\mathrm{c}} \mathrm{d}\tau,$$

$$W(T_{\mathrm{s}}) = \int_0^{T_{\mathrm{s}}} e^{A_{\mathrm{c}}\tau} W_{\mathrm{c}} e^{A_{\mathrm{c}}^{\mathsf{T}}\tau} \mathrm{d}\tau,$$

$$Q(T_{\mathrm{s}}) = \int_0^{T_{\mathrm{s}}} e^{A_{\mathrm{c}}^{\mathsf{T}}\tau} Q_{\mathrm{c}} e^{A_{\mathrm{c}}\tau} \mathrm{d}\tau,$$

$$H(T_{\mathrm{s}}) = \int_0^{T_{\mathrm{s}}} e^{A_{\mathrm{c}}^{\mathsf{T}}\tau} Q_{\mathrm{c}} \int_0^{\tau} e^{A_{\mathrm{c}}s} \mathrm{d}s B_{\mathrm{c}} \mathrm{d}\tau,$$

$$R(T_{\mathrm{s}}) = T_{\mathrm{s}} R_{\mathrm{c}} + B_{\mathrm{c}}^{\mathsf{T}} \int_0^{T_{\mathrm{s}}} \int_0^{\tau} e^{A_{\mathrm{c}}^{\mathsf{T}}s} \mathrm{d}s Q_{\mathrm{c}} \int_0^{\tau} e^{A_{\mathrm{c}}s} \mathrm{d}s \mathrm{d}\tau B_{\mathrm{c}}.$$

Instead of finding a controller that minimizes the cost (5.2) of the continuous-time system (5.1), the goal is now to find a controller that minimizes the cost (5.6) of the discrete-time system (5.5).

5.2.2. Loss and Delay seen by the Controller

When the controller and the routing within the communication system are jointly designed, the arrival probabilities seen by the controller and the sampling time follow from the loss and delay of the communication system and the routing. Within a networked control system, there are three overlay nodes: the sensor, the controller, and the actuator. Thus, we use the vectors $\mathbf{z}_{\mathrm{pos,s}}$, $\mathbf{z}_{\mathrm{pos,c}}$, and $\mathbf{z}_{\mathrm{pos,a}}$ to represent their position. Moreover, we also have to send measurement packets from the sensor to the controller, control packets from the controller to the actuator, and acknowledgments from the actuator to the controller. We use the vectors $\mathbf{z}_{\mathrm{sc}}$, $\mathbf{z}_{\mathrm{ca}}$, and $\mathbf{z}_{\mathrm{ac}}$ for representing these routes.

From the point of view of the controller, the loss along a route can be modeled as a Bernoulli loss process. Thus, we use $\gamma_k \in \{0,1\}$ with $\Pr\{\gamma_k = 1\} = \bar{p}_{\mathrm{sc}}$ to describe the loss between sensor and controller, $\beta_k \in \{0,1\}$ with $\Pr\{\beta_k = 1\} = \bar{p}_{\mathrm{ca}}$ to describe the loss between controller and actuator, and $\mathrm{ACK}_k \in \{0,1\}$ with $\Pr\{\mathrm{ACK}_k = 1\} = \bar{p}_{\mathrm{ac}}$ for the loss between actuator and controller. Similarly, $\bar{d}_{\mathrm{sc}}$, $\bar{d}_{\mathrm{ca}}$, and $\bar{d}_{\mathrm{ac}}$ are used to denote the corresponding delay.

Since the focus of this chapter is on the design on the network layer, we assume that lost packets are not retransmitted, i.e., we only consider the standard fast sampling scheme of the previous chapter, depicted in Figure 5.2. As already stated, we moreover assume that the actuator and sensor are synchronized. The controller waits long enough to receive the measurement and the acknowledgement before calculating the next control input. Thus, the minimal sampling time is

$$T_{\mathrm{S,min}} = \max\{\bar{d}_{\mathrm{sc}}, \bar{d}_{\mathrm{ac}}\} + \bar{d}_{\mathrm{ca}}.$$

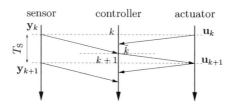

Figure 5.2.: The standard fast timing.

5.2.3. Controller Design

As in the previous chapter, the input to the plant is set to zero when the control packet is missing. Thus, we again use $\mathbf{u}_k^c$ for the control input calculated by the controller. From the point of view of the controller, the state of the plant evolves as

$$\mathbf{x}_{k+1} = A(T_\mathrm{S})\mathbf{x}_k + \beta_k B(T_\mathrm{S})\mathbf{u}_k^c + \mathbf{w}_k, \tag{5.7}$$

the measurements received by the controller are

$$\mathbf{y}_k^c = \begin{cases} C\mathbf{x}_k + \mathbf{v}_k, & \gamma_k = 1, \\ \emptyset, & \gamma_k = 0, \end{cases} \tag{5.8}$$

and the cost becomes

$$J(T_\mathrm{S}) = \mathbf{x}_N^\mathsf{T} F \mathbf{x}_N + \sum_{k=0}^{N} c_k(T_\mathrm{S}), \tag{5.9}$$

$$c_k(T_\mathrm{S}) = \mathbf{x}_k^\mathsf{T} Q(T_\mathrm{S})\mathbf{x}_k + 2\beta_k \mathbf{x}_k^\mathsf{T} H(T_\mathrm{S})\mathbf{u}_k^c + \beta_k \mathbf{u}_k^{c\,\mathsf{T}} R(T_\mathrm{S})\mathbf{u}_k^c.$$

In Schenato et al. (2007), it is shown that the separation principle does not hold and the optimal control law is nonlinear when the acknowledgments of the control packets are not reliable. For simplicity, we hence restrict ourselves to linear controllers with constant gains L and K:

$$\hat{\mathbf{x}}_{k+1} = A\hat{\mathbf{x}}_k + \mathrm{E}[\beta_k|\mathrm{ACK}_k]B\mathbf{u}_k^c + \gamma_k L(\mathbf{y}_k^c - C\hat{\mathbf{x}}_k), \tag{5.10a}$$

$$\mathbf{u}_k^c = -K\hat{\mathbf{x}}_k, \tag{5.10b}$$

where

$$\mathrm{E}[\beta_k|\mathrm{ACK}_k] = \begin{cases} 1 & \text{if } \mathrm{ACK}_k = 1 \\ \bar{\epsilon} & \text{if } \mathrm{ACK}_k = 0 \end{cases}, \qquad \bar{\epsilon} = \frac{\overline{p}_\mathrm{ca}(1 - \overline{p}_\mathrm{ac})}{1 - \overline{p}_\mathrm{ca}\overline{p}_\mathrm{ac}}.$$

Theorem 5.1. *Suppose, controller* (5.10) *is used to control system* (5.7) *with measurements* (5.8). *Moreover the cost is given by* (5.9). *Then, for* $N \to \infty$, *the expected cost per step is minimized for*

$$L = A\overline{P}C^\mathsf{T}\left(C\overline{P}C^\mathsf{T} + V\right)^{-1} \tag{5.11}$$

$$K = \left(B^\mathsf{T}\overline{\Lambda}B + R + \phi B^\mathsf{T}\underline{\Lambda}B\right)^{-1}\left(B^\mathsf{T}\overline{\Lambda}A + H^\mathsf{T}\right), \tag{5.12}$$

where

$$\overline{P} = A\overline{P}A^\mathsf{T} + W - \overline{p}_{sc}L(C\overline{P}C^\mathsf{T} + V)L^\mathsf{T}$$
$$+ \overline{p}_{ca}\phi BK\underline{P}K^\mathsf{T}B^\mathsf{T} \tag{5.13}$$
$$\underline{P} = A\underline{P}A^\mathsf{T} + \overline{p}_{sc}L(C\overline{P}C^\mathsf{T} + V)L^\mathsf{T}$$
$$- \overline{p}_{ca}(BK\underline{P}A^\mathsf{T} + A\underline{P}K^\mathsf{T}B^\mathsf{T}) + \sigma BK\underline{P}K^\mathsf{T}B^\mathsf{T} \tag{5.14}$$
$$\overline{\Lambda} = A^\mathsf{T}\overline{\Lambda}A + Q - \overline{p}_{ca}K^\mathsf{T}(B^\mathsf{T}\overline{\Lambda}B + R + \phi B^\mathsf{T}\underline{\Lambda}B)K \tag{5.15}$$
$$\underline{\Lambda} = A^\mathsf{T}\underline{\Lambda}A + \overline{p}_{sc}(C^\mathsf{T}L^\mathsf{T}\underline{\Lambda}LC - C^\mathsf{T}L^\mathsf{T}\underline{\Lambda}A - A^\mathsf{T}\underline{\Lambda}LC)$$
$$+ \overline{p}_{ca}K^\mathsf{T}(B^\mathsf{T}\overline{\Lambda}B + R + \phi B^\mathsf{T}\underline{\Lambda}B)K \tag{5.16}$$

with

$$\phi = \frac{(1 - \overline{p}_{ca})(1 - \overline{p}_{ac})}{1 - \overline{p}_{ca}\overline{p}_{ac}},$$
$$\sigma = \overline{p}_{ca}\overline{p}_{ac} + \frac{\overline{p}_{ca}^2(1 - \overline{p}_{ac})^2}{1 - \overline{p}_{ca}\overline{p}_{ac}}.$$

The expected cost per step is

$$\mathrm{E}[c_\infty] = \mathrm{Tr}\left(\begin{bmatrix} Q & -\overline{p}_{ca}HK \\ -\overline{p}_{ca}K^\mathsf{T}H^\mathsf{T} & \overline{p}_{ca}K^\mathsf{T}RK \end{bmatrix}\begin{bmatrix} \underline{P} + \overline{P} & \underline{P} \\ \underline{P} & \underline{P} \end{bmatrix}\right).$$

The proof of Theorem 5.1 is given in Section C.1 of the appendix.

When control or acknowledgement packets can get lost, the control input is not known exactly by the controller. Moreover, note that we have $\phi \neq 0$ in this case. Consequently, (5.13) - (5.16) are coupled. Furthermore, the controller gain K and the estimator gain L can not be designed separately.

On the other hand, when the control input is exactly known by the controller, which is the case when all control packets arrive or all acknowledgement packets arrive, then $\phi = 0$. In this case, the controller design problem becomes much simpler. Therefore, first note that (5.12) simplifies to

$$K = (B^\mathsf{T}\overline{\Lambda}B + R)^{-1}(B^\mathsf{T}\overline{\Lambda}A + H^\mathsf{T}),$$

i.e., it is independent of $\underline{\Lambda}$. Moreover, (5.13) and (5.15) simplify to

$$\overline{P} = A\overline{P}A^\mathsf{T} + W - \overline{p}_{sc}L(C\overline{P}C^\mathsf{T} + V)L^\mathsf{T},$$
$$\overline{\Lambda} = A^\mathsf{T}\overline{\Lambda}A + Q - \overline{p}_{ca}K^\mathsf{T}(B^\mathsf{T}\overline{\Lambda}B + R).$$

I.e., $\overline{P}$, no longer depends on $\underline{P}$. Similarly, $\overline{\Lambda}$ no longer depends on $\underline{\Lambda}$. Consequently, the controller gain K and the estimator gain L can be obtained from the following two modified algebraic Riccati equations.

$$L = A\overline{P}C^\mathsf{T}(C\overline{P}C^\mathsf{T} + V)^{-1}$$
$$\overline{P} = A\overline{P}A^\mathsf{T} + W - \overline{p}_{sc}L(C\overline{P}C^\mathsf{T} + V)L^\mathsf{T}$$

and

$$K = \left(B^\mathsf{T}\overline{\Lambda}B + R\right)^{-1}\left(B^\mathsf{T}\overline{\Lambda}A + H^\mathsf{T}\right),$$
$$\overline{\Lambda} = A^\mathsf{T}\overline{\Lambda}A + Q - \overline{p}_{\mathrm{ca}}K^\mathsf{T}\left(B^\mathsf{T}\overline{\Lambda}B + R\right).$$

Note that these Riccati equations are no longer coupled, i.e., the controller gain K and the estimator gain L can be designed separately.

Note that Theorem 5.1 gives the cost per step, but the step size is given by the sampling time T_S. As in the previous chapter, we hence use the relative cost per step $\tilde{c}_\infty = \mathrm{E}[c_\infty]/T_\mathrm{S}$ to compare different configurations fairly. Moreover, since $\tilde{c}_\infty$ depends on the end-to-end delays and arrival probabilities, we write $\tilde{c}_\infty(\overline{d}_\mathrm{sc}, \overline{d}_\mathrm{ca}, \overline{d}_\mathrm{ac}, \overline{p}_\mathrm{sc}, \overline{p}_\mathrm{ca}, \overline{p}_\mathrm{ac})$ in the following to explicitly express this dependency.

Assuming that the position of the sensor and actuator are fixed but the controller can be placed freely, the optimization problem becomes

$$\begin{aligned}
\min \quad & \tilde{c}_\infty(\overline{d}_\mathrm{sc}, \overline{d}_\mathrm{ca}, \overline{d}_\mathrm{ac}, \overline{p}_\mathrm{sc}, \overline{p}_\mathrm{ca}, \overline{p}_\mathrm{ac}) \\
\text{s.t.} \quad & E\mathbf{z}_\mathrm{sc} = \mathbf{z}_\mathrm{pos,s} - \mathbf{z}_\mathrm{pos,c}, \\
& E\mathbf{z}_\mathrm{ca} = \mathbf{z}_\mathrm{pos,c} - \mathbf{z}_\mathrm{pos,a}, \\
& E\mathbf{z}_\mathrm{ac} = \mathbf{z}_\mathrm{pos,a} - \mathbf{z}_\mathrm{pos,c}, \\
& \mathbf{z}_\mathrm{sc}, \mathbf{z}_\mathrm{ca}, \mathbf{z}_\mathrm{ac} \in \{0,1\}^{|\mathcal{E}|}, \\
& \mathbf{z}_\mathrm{pos,c} \in \{0,1\}^{|\mathcal{N}|}, \mathbf{1}^\mathsf{T}\mathbf{z}_\mathrm{pos,c} = 1,
\end{aligned}$$

with

$$\begin{array}{lll}
\overline{d}_\mathrm{sc} = \underline{\mathbf{d}}^\mathsf{T}\mathbf{z}_\mathrm{sc}, & \overline{d}_\mathrm{ca} = \underline{\mathbf{d}}^\mathsf{T}\mathbf{z}_\mathrm{ca}, & \overline{d}_\mathrm{ac} = \underline{\mathbf{d}}^\mathsf{T}\mathbf{z}_\mathrm{ca}, \\
\log(\overline{p}_\mathrm{sc}) = \underline{\tilde{\mathbf{p}}}^\mathsf{T}\mathbf{z}_\mathrm{sc}, & \log(\overline{p}_\mathrm{ca}) = \underline{\tilde{\mathbf{p}}}^\mathsf{T}\mathbf{z}_\mathrm{ca}, & \log(\overline{p}_\mathrm{ac}) = \underline{\tilde{\mathbf{p}}}^\mathsf{T}\mathbf{z}_\mathrm{ac}.
\end{array}$$

Unfortunately, this optimization problem is hard to solve due to its nonlinear cost function and integer constraints. However, in Carabelli et al. (2012), it is shown that for a linear cost function, this optimization problem can be reformulated as a linear problem by relaxing the integer constraints.

5.3. Examples

In this section, we demonstrate two interesting effects that appear when the network layer is optimized. Therefore, we use the same system as already considered in Example 3 of the previous chapter but with an output $\mathbf{y} = C_\mathrm{c}\mathbf{x} + \mathbf{v}$ and two different underlay networks. For both examples, the matrices of system (5.1) are

$$A_\mathrm{c} = \begin{bmatrix} 0 & 0.5 \\ -0.5 & 0.5 \end{bmatrix}, \quad B_\mathrm{c} = \begin{bmatrix} 0 \\ 1 \end{bmatrix}, \quad C_\mathrm{c} = \begin{bmatrix} 0 & 1 \end{bmatrix},$$

$$W_\mathrm{c} = \begin{bmatrix} 1 & 0 \\ 0 & 1 \end{bmatrix}, \quad V_\mathrm{c} = 1, \quad Q_\mathrm{c} = \begin{bmatrix} 1 & 0 \\ 0 & 1 \end{bmatrix}, \quad R_\mathrm{c} = 1.$$

Example 1

Figure 5.3 shows the setup of the first example. The loss and delay of the different links are

$$\underline{p}_0 = 0.99, \qquad \underline{p}_1 = 0.6, \qquad \underline{p}_2 = 0.7, \qquad \underline{p}_3 = 0.99, \qquad \underline{p}_4 = 0.95,$$
$$\underline{d}_0 = 0.01, \qquad \underline{d}_1 = 0.01, \qquad \underline{d}_2 = 0.1, \qquad \underline{d}_3 = 0.1, \qquad \underline{d}_4 = 0.03.$$

Note that there are four different routes between the sensor and the actuator:

$$\text{route 1:} \quad 1 \to 2 \to 3 \to 4 \to 10 \to 11,$$
$$\text{route 2:} \quad 1 \to 2 \to 5 \to 10 \to 11,$$
$$\text{route 3:} \quad 1 \to 2 \to 6 \to 7 \to 10 \to 11,$$
$$\text{route 4:} \quad 1 \to 2 \to 8 \to 9 \to 10 \to 11.$$

It is easy to see that route 1 is the route with the shortest delay, route 2 the one with the minimal number of nodes (hops), and route 3 the one with the highest arrival probability. Since there are several routes between sensor and actuator the interesting questions are: Along which route do we get the best performance? Which controller position gives the best performance?

Figure 5.4 shows the relative cost per step for the different positions of the controller on the different routes. The optimal controller position is node 11 (actuator node) but the optimal route is neither the route with the minimal delay, the route with the minimal number of nodes (hops), nor the route with the highest arrival probability. Instead, it is a route with a good compromise between loss and delay.

Interestingly, when the packets are routed along route 2, the best controller position is not at node 11 (actuator node), where we get $\tilde{J} = 15.082$, but at node 10, where we get $\tilde{J} = 14.930$. From this example, we see that the conclusion of Robinson and Kumar (2008), i.e., that for full state measurements, the best performance is achieved when the controller is at the same node as the actuator, does not hold for general measurements. Hence, we study this effect with the help of a line graph within the next example in some more detail.

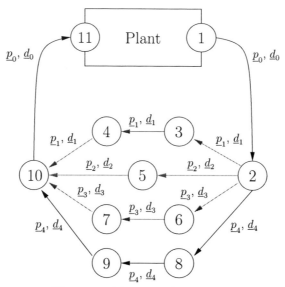

Figure 5.3.: The setup of Example 1.

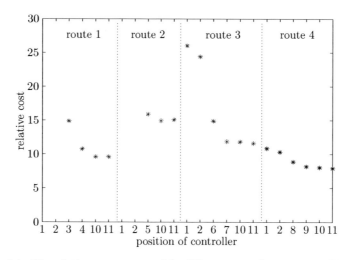

Figure 5.4.: The relative cost per step of the different controller positions of Example 1.

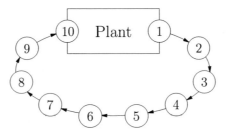

Figure 5.5.: The setup of Example 2.

Example 2

In this example, the communication network is a line graph with 10 nodes, as depicted in Figure 5.5. The position of the controller is not fixed and should be placed such that the relative cost per step is minimized. Since the graph is a line graph, the loss and delay between sensor, controller, and actuator are simple to derive for a given position of the controller. To see how the control performance depends on the arrival probability and the position of the controller, we assume $\underline{d} = 0.01$ on all links and vary the arrival probabilities of the links and calculate the relative cost per step for each position of the controller.

Figure 5.6a shows how the delay between the sensor, controller, and actuator and the resulting sampling time depends on the position of the controller. When the controller is placed closer to the sensor than to the actuator, the sampling time is twice the delay between controller and actuator. When the controller is placed closer to the actuator than to the sensor, the sampling time is just the delay between the sensor and the actuator.

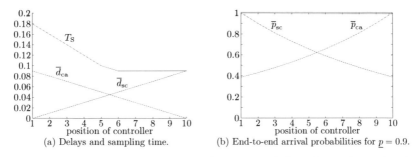

(a) Delays and sampling time.
(b) End-to-end arrival probabilities for $\underline{p} = 0.9$.

Figure 5.6.: The delays, the resulting sampling time, and the end-to-end arrival probability.

Figure 5.6b shows how the end-to-end arrival probabilities depend on the position of the controller for $\underline{p} = 0.9$. Obviously, when the controller is placed close to the sensor, the arrival probability of measurement packets is relatively high but the arrival probability of control and acknowledgement packets is relatively low. Similarly, when the controller is placed close to the actuator, the arrival probability of control and acknowledgement packets is relatively high but the arrival probability of measurement packets is relatively small. Since all packets are important, a compromise must be found.

Finally, Figure 5.7 shows how the relative cost per step depends on the position of the controller for different arrival probabilities of the links. For $\underline{p} = 0.95$, a stabilizing controller can be found for all positions of the controller; the minimal relative cost per step is achieved when the controller is located at node 10, i.e., at the actuator. For $\underline{p} = 0.9$, no stabilizing controller can be found when the controller is located at node 1, i.e., at the sensor. Again, the minimal relative cost per step is achieved when the controller is located at the actuator. Interestingly, placing the controller at node 10 is no longer optimal for $\underline{p} = 0.85$. Here, the relative cost per step is $\tilde{c}_\infty = 12.326$ when the controller is located at node 9 and $\tilde{c}_\infty = 12.691$ when the controller is located at node 10. Moreover, for $\underline{p} = 0.8$, no stabilizing controller can be found when the controller is placed at node 10 and the minimal relative cost per step is achieved when the controller is located at node 7. Finally, for $\underline{p} = 0.725$, a stabilizing controller can only be found when the controller is located at node 6.

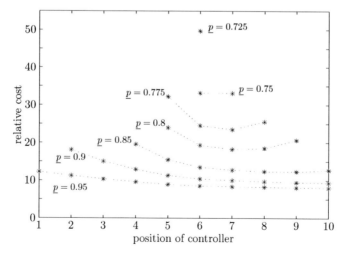

Figure 5.7.: The relative cost per step of the different controller positions in Example 2.

5.4. Summary

In this chapter, we considered the joint design of controller and routing through the communication system. Therefore, we modeled the underlay communication system as a graph and showed how the end-to-end loss and delay can be derived from the loss and delay of the underlay links and the route. Based on these results, we wrote the joint optimal control and routing problem as an optimization problem with integer constraints. With the help of two examples, we finally showed some interesting effects that appear when the controller and routing are considered together. These examples showed that the optimal route is not the one with the shortest delay, highest arrival probability, or minimal number of hops. Moreover, when the state can not be measured directly, the optimal position of the controller is not always the actuator node, as it is the case when the state can be measured directly.

Chapter 6.

Optimization within the MAC Layer: Time-Triggered vs. Event-Based Control

In Chapter 4, we showed that it is possible to trade loss against a longer sampling time by retransmitting lost packets. Thereby, we avoided to increase the packet rate because loss and delay of a link generally depend on its usage. Within this chapter, we study in detail how this effect influences the design of a networked control system and the achievable performance by analyzing the interaction between control and communication within the MAC layer. Considering the details of the medium access allows us to use more accurate models of the communication system, where the loss probability and delay explicitly depend on the traffic pattern and network load, which are determined by the controller design. In doing so, it becomes possible to study how the network load and traffic pattern affect the control performance. This chapter is based on Blind and Allgöwer (2013b); earlier versions are published in parts in Blind and Allgöwer (2011a,b,c).

6.1. Introduction

In the field of communication systems, it is well known that loss and delay depend on the communication protocol, network resources, their utilization, and other factors. However, in most works in the field of networked control systems, loss and delay are assumed to be fixed in the sense that they are independent of the usage of the communication system. One example of this gap is the question how to proceed after a packet loss. In Mesquita et al. (2012), it is suggested to increase the number of transmitted packets after a packet loss. This is in sharp contrast to the congestion avoidance algorithm of TCP, where the offered load is reduced after a packet loss, as first described in Jacobson (1988).

In this chapter, we go one step in the direction of closing this gap between control and communication theory by taking the medium access into account when comparing time-triggered and event-based control over a shared communication system. The considered problem setup is depicted in Figure 6.1. There are N plants; each plant

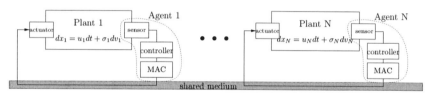

Figure 6.1.: The considered setup.

is an integrator system with noise, which is controlled by an agent. Each agent has to sample the plant and calculate the control input. Moreover, since all agents use the same shared medium, they are also responsible for the medium access. To study the interaction between control and communication, we use mathematical models of different MAC protocols, that model the dependency of loss and delay on the network load and traffic pattern. Since the network load and traffic pattern are both affected by the controller design, we consider only loss and delay from the MAC layer and assume that there are no other sources for packet loss like random noise.

Our detailed analysis of time-triggered and event-based control over a shared communication system is also motivated by the comparison of time-triggered and event-based control in Åström and Bernhardsson (2002), which shows that event-based control gives the same performance with fewer events when the communication is ideal. Consequently event-based control seems to be better suited for networked control systems, see, e.g., Anta and Tabuada (2010); Heemels et al. (2008); Henningsson et al. (2008); Kofman and Braslavsky (2006); Lunze and Lehmann (2010); Mazo and Tabuada (2011); Tabuada (2007); Wang and Lemmon (2008, 2011). However, it is not that simple when the medium access is taken into account. When using time-triggered control, all packet generation times are known in advance. In contrast, when using event-based control, the packet generation times depend on the state of the system and are thus not known in advance. Consequently, when comparing time-triggered and event-based control, this difference in the traffic pattern and its effect on the medium access must be taken into account. Hence, in this chapter, time-triggered control with the two most well known deterministic communication protocols, Time Division Multiple Access (TDMA) and Frequency Division Multiple Access (FDMA) is compared to event-based control with different contention based communication protocols: pure ALOHA, slotted ALOHA, a queueing system, and Erlang's loss model. The result of this comparison is summarized in Table 6.1 and depicted in Figure 6.9. It turns out that time-triggered control with either TDMA or FDMA outperforms event-based control with pure or slotted ALOHA. However, event-based control with a properly designed queueing system gives an even better performance.

Event-based control with a shared medium is also studied in Cervin and Henningsson (2008) and Henningsson and Cervin (2010). A simulation based approach is used in Cervin and Henningsson (2008) to compare the performance of time-triggered and event-based control with different communication protocols. It is concluded that event-based control with a CSMA (Carrier Sense Multiple Access) protocol gives the best performance. In Henningsson and Cervin (2010), the impulsive control of an integrator system with a CSMA communication protocol is considered analytically, as in this chapter. However, there are some important differences. First, the effect of an arriving packet is different. Within this thesis, there is a sharp separation between control and communication. The control input is calculated and passed to the communication system, which sends a packet to the actuator. When the packet arrives, it will be outdated. In contrast, in Henningsson and Cervin (2010) it is assumed that a packet contains up to date information. Moreover, our CSMA model differs from the one in Henningsson and Cervin (2010) by the choice of the next packet after a busy period. In Henningsson and Cervin (2010), it is assumed that the next packet is chosen randomly from the set of packets that are waiting to be transmitted. We assume that the packets are buffered in a global FIFO (first in first out) queue, i.e., they are transmitted in the same order as they arrive. Interestingly, for a large number of agents, we get the same performance for event-based control with CSMA but different sending rates. Finally, in Henningsson and Cervin (2010) the interevent times are assumed to be generated by a Poisson process. Within this chapter, Theorem 6.9 shows that for our event-based control scheme, the interevent times of all agents together converge to a Poisson process as the number of agents approaches infinity. However, this is due to our specific choice of the control and communication scheme and does not necessarily hold for the setup of Henningsson and Cervin (2010).

There also exist more approaches to take the medium access in networked control systems into account. One approach is to give access to the agent with the largest error, see, e.g., Hristu-Varsakelis and Kumar (2002); Walsh et al. (1999). Obviously, this requires a scheme to compare these errors. Another approach is to let the agents transmit only with some probability, see, e.g., Liu and Goldsmith (2004a); Rabi and Stabellini (2008); Zhang (2003). As already stated, it is also possible that the communication system chooses the next agent that is allowed to transmit randomly, see, e.g., Cervin and Henningsson (2008); Henningsson and Cervin (2010); Molin and Hirche (2011). Finally, in Ramesh et al. (2012a,b) the contention resolution mechanism is modeled by a Markov chain. However, except from Ramesh et al. (2012a,b, 2013), these models are difficult to realize or do not represent realistic MAC protocols.

Notation

To denote the sampling time or expected interevent time of agent i, we use T_i. For simplicity, we also use the sending rate $\lambda_i := 1/T_i$. In doing so, the sending rate of a set of agents is just the sum of the individual sending rates. Moreover, the load of an agent is defined as $\rho_i := \tau\lambda_i$, where τ is the packet duration. Again, the load of a set of agents is just the sum of the individual loads.

When considering a function with a parameter, we write $f(x|\lambda)$ to indicate that x is the argument and λ the parameter. Moreover, when considering random variables, a lowercase letter is used for its Probability Density Function (PDF), i.e., $f(x|\lambda)$ and the corresponding uppercase letter for its Cumulative Distribution Function (CDF), i.e., $F(x|\lambda) := \int_0^x f(t|\lambda)\mathrm{d}t$.

Next, we shortly discuss the scale parameter of a family of distributions. Given a random variable z with PDF $f_z(z)$ and CDF $F_z(z)$, we can define a new random variable $x := az$ with PDF $f_x(x|a) = \frac{1}{a}f_z\left(\frac{x}{a}\right)$ and CDF $F_x(x|a) = F_z\left(\frac{x}{a}\right)$. Since the parameter a scales the random variable, it is consequently called *scale parameter*. Moreover, the inverse of the scale parameter is called *rate parameter*. A more formal and detailed definition can be found in Ferguson (1962) and Section 6.5.1 of Mukhopadhyay (2000), which is summarized in Section B of the Appendix.

6.2. Problem Setup

The control problem of an individual agent is similar to the one in Åström and Bernhardsson (2002); Rabi and Johansson (2009), i.e., there are N integrator systems

$$\mathrm{d}x_i = u_i\mathrm{d}t + \sigma_i\mathrm{d}v_i, \quad i = 1, \dots, N \tag{6.1}$$

where $x_i(t) \in \mathbb{R}$ is the state of system i, the disturbance $v_i(t) \in \mathbb{R}$ is a Wiener process and $u_i(t) \in \mathbb{R}$ the control signal. Note that in Åström and Bernhardsson (2002); Rabi and Johansson (2009) it is assumed that $\sigma_i = 1$.

The control input u_i is a sequence of impulses

$$u_i(t) = \sum_{k \in \mathcal{A}_i} -\delta(t - t_{i,k} - d_{i,k})x_i(t_{i,k}), \tag{6.2}$$

where $t_{i,k}$ is the time of the k-th event generation, $d_{i,k}$ the delay of the corresponding packet, and $\mathcal{A}_i$ the index set of arrived packets. Note that each impulse is such that it resets the state to the origin when applied immediately, i.e., when $d_{i,k} = 0$. Between the events, the input is zero and the system just integrates the noise. Combining (6.1) and (6.2), we get

$$\begin{aligned} x_i^+(t_{i,k} + d_{i,k}) &= x_i(t_{i,k} + d_{i,k}) - x_i(t_{i,k}) && \text{for } t = t_{i,k} + d_{i,k}, k \in \mathcal{A}_i, \\ \mathrm{d}x_i &= \sigma_i\mathrm{d}v_i, \ u_i = 0 && \text{else,} \end{aligned}$$

where x_i^+ is the state of the system directly after applying the impulse.

Obviously, an impulsive control input is not realistic when a physical system is controlled but a good starting point for analyzing the interaction between control and communication. Moreover, when the state of the system represents the estimation error, the estimation error will be zero when a packet arrives that contains the correct measurement.

As in Åström and Bernhardsson (2002); Rabi and Johansson (2009), the variance of the state is used as cost to compare the different control and communication strategies:

$$J_i = \lim_{M \to \infty} \sup \frac{1}{M} \int_0^M \mathrm{E}[x_i(t)^2]\mathrm{d}t. \tag{6.3}$$

In Section 6.3.1 and 6.3.2 we show how the cost depends on the sampling strategy and the three parameters network load, loss probability, and delay. In contrast to most previous works, we do not assume that these parameters are independent from each other. Instead, we use mathematical models of different communication protocols to get the relationship between network load, loss probability, and delay. These mathematical models are based on the following two assumptions on the communication system.

Assumption 6.1. *Each packet takes some time to be transmitted, the* packet duration τ, *which is equal for all packets.*

Assumption 6.2. *When two or more agents send at the same time, the packets collide and are lost.*

The communication protocols differ in the approach to avoid and handle collisions, see Section 6.4 for more details. Thus, the end-to-end delay d and the loss probability q are a result of the control and communication strategy and follow from the mathematical models. Hence, loss and delay depend on the choice of the communication protocol, the packet duration τ, the control strategy, and the network load.

6.3. Sampling Schemes

In this section, we introduce the two most common sampling schemes, namely time-triggered and event-based control. We present the basic idea of the two sampling schemes, the resulting interevent time, as well as the cost for controlling system (6.1) with these sampling schemes.

6.3.1. Time-Triggered Control

Currently, the most common sampling strategy is *time-triggered control*, where agent i samples system (6.1) periodically with a constant sampling time $T_{\mathrm{TT},i}$. Thus, the time between events is constant, i.e., $t_{i,k+1} - t_{i,k} = T_{\mathrm{TT},i}$, and the sending rate and load becomes $\lambda_{\mathrm{TT},i} = 1/T_{\mathrm{TT},i}$ and $\rho_{\mathrm{TT},i} = \tau/T_{\mathrm{TT},i}$. The resulting cost is given in the following theorem.

Theorem 6.3. *Suppose, system (6.1) is controlled by an impulsive time-triggered control scheme with sampling time $T_{\mathrm{TT},i}$, a packet loss probability q_i, and an expected delay d_i. Moreover, loss and delay are assumed to be independent from the state of the system. Then the cost is*

$$J_{\mathrm{TT},i} = \sigma_i^2 \left(\frac{T_{\mathrm{TT},i}}{2} + \frac{T_{\mathrm{TT},i} q_i}{(1 - q_i)} + d_i \right). \tag{6.4}$$

The proof is given in Section D.1 of the appendix.

6.3.2. Event-Based Control

An alternative to time-triggered control is *event-based control*, where the state is used to determine the event times and thereby sample only when necessary. The most simple approach is a threshold policy, where an event is generated whenever the normed state exceeds a predefined constant bound Δ_i, i.e., whenever $|x_i| = \Delta_i$. Obviously, this is not always the optimal sampling scheme. E.g., the optimal sampling for the case that only a limited number of samples is allowed during a fixed time interval is studied in Rabi et al. (2012). In this case, the optimal bounds are time varying. Thus, it is not clear whether the simple threshold policy is optimal, especially when also the effects on the communication system are taken into account. Nevertheless, we use a simple threshold policy to generate the events due to the following reasons. When this policy is used, the interevent-times will be independent and identically distributed (iid). An iid arrival process is a standard assumption for the study of communication systems and allows the usage of results from renewal theory. Moreover, the interevent-times and their distribution are known for this policy and this knowledge is necessary for a mathematical analysis of event-based control over a shared communication system. Finally, this simple threshold policy is the most often considered event-based approach.

When packets are delayed or lost, the simple threshold policy becomes slightly more complex. In this case, the bounds must be chosen such that the state x_i is always between a lower and an upper bound and the correct action is performed when the corresponding packet arrives. To achieve this, the bounds are managed by two state variables, the position of the bounds ($\underline{\Delta}_i$ and $\overline{\Delta}_i$) and the corresponding action ($\underline{\xi}_i$ and $\overline{\xi}_i$). These variables are updated as follows:

- state reaches an active bound, i.e., $x_i(t) = \overline{\Delta}_i$ or $x_i(t) = \underline{\Delta}_i$:
  - send packet with action $\xi_{i,k} = \begin{cases} \overline{\xi}_i & \text{if } x_i(t) = \overline{\Delta}_i \\ \underline{\xi}_i & \text{if } x_i(t) = \underline{\Delta}_i \end{cases}$
  - change position of bounds: $\overline{\Delta}_i^+ = x_i(t) + \Delta_i$, $\underline{\Delta}_i^+ = x_i(t) - \Delta_i$
  - change actions: $\overline{\xi}_i^+ = -\Delta_i$, $\underline{\xi}_i^+ = +\Delta_i$
- packet with action $\xi_{i,k}$ is lost:
  - change actions: $\overline{\xi}_i^+ = \overline{\xi}_i + \xi_{i,k}$, $\underline{\xi}_i^+ = \underline{\xi}_i + \xi_{i,k}$
- packet with action $\xi_{i,k}$ arrives:
  - apply action to state: $x_i^+ = x_i + \xi_{i,k}$
  - change positions of bounds: $\overline{\Delta}_i^+ = \overline{\Delta}_i + \xi_{i,k}$, $\underline{\Delta}_i^+ = \underline{\Delta}_i + \xi_{i,k}$

Figure 6.2 shows an example. The system is started at the origin. At time $t_{i,1}$, the upper bound is reached. Thus, a packet with the action $\xi_{i,1} = -\Delta_i$ is sent and the bounds are set to $\underline{\Delta}_i = 0$ and $\overline{\Delta}_i = 2\Delta_i$. At time $t_{i,2}$, the state reaches the upper

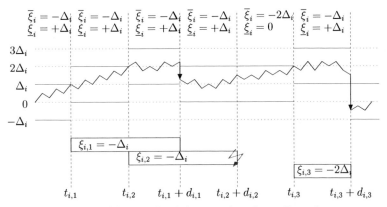

Figure 6.2.: An example of the event-based sampling scheme.

bound $\overline{\Delta}_i$. Again, a packet with the action $\xi_{i,2} = -\Delta_i$ is sent and the bounds are set to $\underline{\Delta}_i = \Delta_i$ and $\overline{\Delta}_i = 3\Delta_i$. At time $t_{i,1} + d_{i,1}$, the first packet arrives. Thus, the state and the bounds are changed by $\xi_{i,1} = -\Delta_i$. At time $t_{i,2} + d_{i,2}$, the second packet is lost. As a consequence thereof, the action is changed to $\underline{\xi}_i = 0$ and $\overline{\xi}_i = -2\Delta_i$. At time $t_{i,3}$, the upper bound is reached. Thus, a packet with the action $\xi_{i,3} = -2\Delta_i$ is sent and the bounds are changed to $\underline{\Delta}_i = \Delta_i$ and $\overline{\Delta}_i = 3\Delta_i$. Moreover, the action is set back to $\underline{\xi}_i = \Delta_i$ and $\overline{\xi}_i = -\Delta_i$. At time $t_{i,3} + d_{i,3}$, the corresponding packet arrives and the state and the bounds are changed by $-2\Delta_i$.

By using this scheme, the state is at the midpoint of the two bounds after every event generation. Moreover, since system (6.1) is an integrator system with noise, its future development is independent from the current state. Thus, the distribution of the interevent times does not depend on the past, i.e., the interevent times are iid. This observation is crucial for the proof of Theorem 6.9 since it allows the usage of results from renewal theory.

Moreover, note that the presented scheme requires that packet loss is realized by the sender. Within the MAC layer, this can be realized by a special jam signal that indicates a packet collision, as done in Ethernet. Another approach would be the usage of a second channel to acknowledge successful transmissions. Since acknowledgements are only sent upon a successful reception, there will be only one sender and thus no arbitration conflicts on this channel.

In Åström and Bernhardsson (2002), the expected interevent time and the cost has been derived for $\sigma_i = 1$ and an ideal communication system. In Rabi and Johansson (2009), these results have been extended to a communication system with packet loss. Moreover, Rabi and Johansson (2009) also gives the probability density function of the interevent times. The following theorems extend the results of Åström and Bernhardsson (2002); Rabi and Johansson (2009) to arbitrary σ_i and delayed events.

Theorem 6.4. *Suppose, system* (6.1) *is controlled by an impulsive event-based control scheme with boundary increment Δ_i. Then, the expected interevent time $T_{\text{EB},i}$, sending rate $\lambda_{\text{EB},i}$ and load $\rho_{\text{EB},i}$ is*

$$T_{\text{EB},i} := \text{E}[t_{i,k+1} - t_{i,k}] = \frac{\Delta_i^2}{\sigma_i^2}, \qquad \lambda_{\text{EB},i} = \frac{\sigma_i^2}{\Delta_i^2}, \qquad \rho_{\text{EB},i} = \frac{\tau \sigma_i^2}{\Delta_i^2}. \tag{6.5}$$

The proof is given in Section D.2 of the appendix.

Theorem 6.5. *Suppose, system* (6.1) *is controlled by an impulsive event-based control scheme with boundary increment Δ_i, a packet loss probability q_i, and an expected delay d_i. Moreover, loss and delay are assumed to be independent from the state of the system. Then the cost is*

$$J_{\text{EB},i} = \frac{\Delta_i^2}{6} + \frac{\Delta_i^2 q_i}{(1 - q_i)} + \sigma_i^2 d_i = \sigma_i^2 \left(\frac{T_{\text{EB},i}}{6} + \frac{T_{\text{EB},i} q_i}{(1 - q_i)} + d_i \right). \tag{6.6}$$

The proof is given in Section D.3 of the appendix.

Remark 6.6. *From* (6.4) *and* (6.6)*, we see that the noise intensity σ_i directly affects the cost. Moreover, for event-based control, we see from* (6.5) *that the offered load ρ_i also depends on the noise intensity σ_i. Since loss and delay depend on the load (see Section 6.4), the noise intensity σ_i affects the cost also indirectly. However, for any given σ_i, the offered load ρ_i can be chosen arbitrarily by a proper choice of the boundary increment Δ_i.*

Lemma 6.7. *Suppose, system* (6.1) *is controlled by an impulsive event-based control scheme with boundary increment Δ_i. Then, the Probability Density Function (PDF) of the interarrival times is*

$$f_{\text{EB}}(t|\lambda_{\text{EB},i}) = \sqrt{\frac{2}{\pi \lambda_{\text{EB},i} t^3}} \sum_{k=-\infty}^{\infty} (4k+1) \exp\left(-\frac{(4k+1)^2}{2\lambda_{\text{EB},i} t} \right). \tag{6.7}$$

Lemma 6.7 follows from *(Feller, 1950, Section 14.9, Problem 7)* by adapting the notation.

Remark 6.8. *The PDF* (6.7) *belongs to a scale family with scale parameter $\lambda_{\text{EB},i}^{-1}$, i.e., the sending rate $\lambda_{\text{EB},i}$ is the rate parameter of this PDF.*

Figure 6.3, shows a plot of the PDF of the normalized interarrival time $f_{\text{EB}}(t|1)$, see Navarro and Fuss (2009) how to compute this PDF fast and accurately. This plot reflects our intuition. For very small values of t, i.e., directly after an event, it is very unlikely that there will be a new event. Similarly, very long interevent times are also unlikely.

Unfortunately, (6.7) is not an integrable series, and thus it is difficult to derive analytical results based in this equation, see also Rabi and Johansson (2009). Nevertheless, it is possible to show that the arrival process of all agents together converges to a Poisson process for $N \to \infty$.

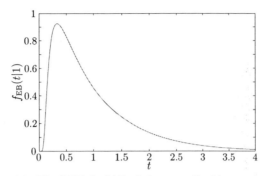

Figure 6.3.: The PDF $f_{\text{EB}}(t|1)$ of the normalized interarrival time.

Theorem 6.9. *Suppose all N agents use the event-based control scheme to control system (6.1). Moreover, all agents send with the same rate λ, such that $\lambda_\Sigma = N\lambda < \infty$. As $N \rightarrow \infty$, the superposition of the arrival processes of all agents approaches a Poisson process with rate λ_Σ.*

The proof is given in Section D.4 of the appendix.

6.3.3. Comparing Time-Triggered and Event-Based Control

Within the field of networked control systems, the communication system is often modeled by some loss probability and/or delay. However, when modeling the communication system with some more details, it turns out that loss and delay depend on the communication system and its usage, i.e., its load and traffic pattern, see, e.g., Rom and Sidi (1990); Tanenbaum (2003). More precisely loss and delay generally increase with the load, i.e., the higher the load, the higher the loss probability and the longer the delay. When taking this into account, the comparison between time-triggered and event-based control becomes very interesting.

First, note that event-based control requires fewer events to achieve the same performance as time-triggered control. This reduced load would lead to a smaller loss probability and a shorter delay, thereby potentially improving the performance. Thus, comparing time-triggered and event-based control under the assumption that loss and delay are equal might be considered as unfair. Obviously, this could be solved by assuming an equal load. Unfortunately, the two sampling schemes generate very different traffic patterns and thus result in different loss probabilities and delays for the same load. For a realistic comparison of time-triggered and event-based control with a shared communication system, we hence assume that the available capacity of the communication system is equal. Packet loss and delay are a consequence of the sampling

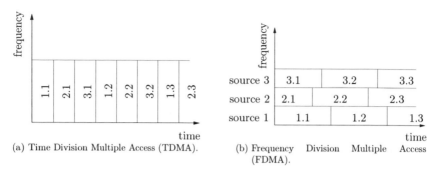

(a) Time Division Multiple Access (TDMA).

(b) Frequency Division Multiple Access (FDMA).

Figure 6.4.: Deterministic protocols.

scheme and medium access. Therefore, the next section contains several mathematical models of a communication system that express how loss and delay depend on the load of the communication system.

6.4. Communication System

In this section, we present several communication protocols that are used to study control and communication in the next section. The considered protocols form the basis of more advanced protocols but are simple enough to be analyzed mathematically. As already stated in Section 6.2, these models are based on the assumption that the packet duration is constant (Assumption 6.1) and the assumption that a packet is lost when two or more agents send simultaneously (Assumption 6.2). The considered protocols differ in the way how collisions are avoided and handled. In the class of *deterministic protocols* collisions are avoided altogether by reserving the resources in advance. Thus, these protocols offer a high Quality of Service (QoS) but are not flexible. On the other hand, in *contention based protocols* the agents are allowed to offer packets at arbitrary times but have to deal with loss and delay caused by arbitration conflicts.

6.4.1. Deterministic MAC Protocols

Figure 6.4 depicts the principle of the two considered deterministic MAC protocols Time Division Multiple Access (TDMA) and Frequency Division Multiple Access (FDMA). As depicted in Figure 6.4a, each agent gets the full bandwidth for a short time when TDMA is used as MAC protocol. In contrast, each agent gets a part of the bandwidth for the full time when FDMA is used, see Figure 6.4b.

Time Division Multiple Access (TDMA)

In TDMA, the sending times are assigned in advance, thereby avoiding collisions and minimizing the waiting time. In doing so, the delay is just the packet duration and there is no loss, i.e.,

$$d_{\text{TDMA},i} = \tau, \tag{6.8}$$

$$q_{\text{TDMA},i} = 0. \tag{6.9}$$

Using TDMA, the network load can not exceed one, i.e.,

$$\rho_\Sigma \le 1 \tag{6.10}$$

must be guaranteed during the assignment of the sending times.

Frequency Division Multiple Access (FDMA)

Here, the available bandwidth is divided between the agents. Agent i gets $1/m_i$-th of the available bandwidth, where m_i must be such that $\sum 1/m_i \le 1$. Since an agent gets $1/m_i$-th of the bandwidth, the time to transmit a packet becomes m_i times larger. Thus, the delay is

$$d_{\text{FDMA},i} = m_i \tau. \tag{6.11}$$

Moreover, since each agent sends on its own frequency band, collisions will not occur and thus there are no losses, i.e.,

$$q_{\text{FDMA},i} = 0. \tag{6.12}$$

Obviously, each agent can send only one packet at the time, so the sampling times must be chosen such that $T_i > m_i \tau = d_i$, i.e., its load must fulfill $\rho_i < 1/m_i$. As a consequence thereof, the network load can not exceed one, i.e.,

$$\rho_\Sigma \le 1 \tag{6.13}$$

must hold.

6.4.2. Contention Based MAC Protocols

When the resources are not reserved in advance but are assigned dynamically on request, the most simple approach is to allow the agents to start sending whenever they have something to send and accept collisions as done in pure ALOHA. Since this results in many collisions, there exist many approaches to avoid them. In slotted ALOHA, the times an agent is allowed to start sending is restricted. In the class of CSMA protocols, the agents sense the medium before they start to send. When the medium is found idle, the agent immediately starts to send. The CSMA protocols differ when the medium is found busy. One approach, called one-persistent CSMA,

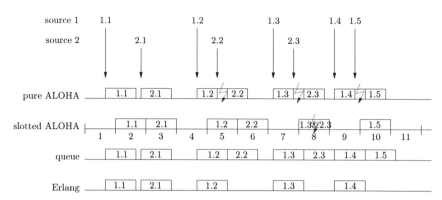

Figure 6.5.: Contention based protocols.

is to wait until the medium becomes idle and then start to send. This approach has the drawback that there might be several agents waiting until the medium becomes idle and then simultaneously start to send and thus collide. We use a variation of this approach in Section 6.4.2, where we assume that a packet is buffered in a global FIFO queue when the medium is found busy. Another approach is to schedule the packet for retransmission at some later time (non-persistent CSMA) or drop the packet (Erlang's loss model). Figure 6.5 depicts an example of the medium access for the considered contention based MAC protocols. These examples are explained in the corresponding sections.

It remains to answer the question what to do when a packet is not successfully transmitted due to a collision or a busy medium. There exist two well known approaches:

Scheme 1: Give up, i.e., drop the packet and accept its loss.

Scheme 2: Schedule the packet for retransmission at some later time.

Note that only Scheme 2 results in a reliable communication, as often required for classic communication tasks, like file transfer. However, in the field of communication theory, it is a well known fact, that retransmitting packets might lead to instability of the communication system. To solve this problem, a backoff algorithm, which delays retransmissions, can be used. In Kleinrock and Lam (1973) a slotted ALOHA system is studied and it is shown that the throughput increases with the retransmission time; it is also shown that the throughput predicted by the Poisson assumption is only approached when the retransmission time approaches infinity. In Lam and Kleinrock (1975a,b), it is suggested to increase the retransmission time with the number of previous losses, as done in Ethernet. However, even with the exponential backoff algorithm used in Ethernet, the communication system might still be unstable, see

Aldous (1987); Kelly (1985). To sum up, when retransmitting packets, a too short retransmission time must be avoided.

Moreover, for real time data, Scheme 2 might not be the best approach since it requires to send outdated packets. For the considered setup both approaches can be compared analytically to answer the question whether to retransmit or drop packets that could not be transmitted successfully.

Theorem 6.10. *Suppose that both, the probability of an unsuccessful transmission (due to a collision or a busy medium) and the expected delay, are increasing with the network load. Moreover, either the probability of an unsuccessful transmission or the expected delay, or both, is strictly increasing with the network load. Furthermore, when using Scheme 2, the expected retransmission time is the same as the expected interevent time. Then, Scheme 1 gives a better performance than Scheme 2.*

The proof is given in Section D.5 of the appendix.

From Theorem 6.10, we see that it is better to accept a packet as lost and let the underlying event-based controller generate the next packet than retransmitting this packet. Consequently, we assume that all packets, that are not transmitted successfully, are dropped and not retransmitted.

Note that there is an important difference between retransmitting packets and buffering them in a global queue. When the packet is retransmitted by the agent after some time, the agent chooses this time, waits until it has past, and then tries to send the packet again. In contrast, when the packets are buffered in a global FIFO queue, packets are transmitted in the order they arrive at the queue. When a packet is enqueued, its delay is the transmission time of all packets currently in the queue.

Pure ALOHA

In pure ALOHA, each agent starts to send whenever it has something to send; packets collide and are lost when two or more agents send simultaneously, see, e.g., Abramson (1970); Rom and Sidi (1990); Tanenbaum (2003). Note that an agent that starts to send might destroy a packet that is already partly transmitted. As a consequence thereof, the vulnerable period[1] is twice the packet duration τ. Although, this protocol is relatively old, it is still used since it is not always possible to sense the medium, see Abramson (2009).

Figure 6.5 shows an example of the medium access. Here, packet 1.2 and 2.2 are both lost when source 2 starts to send packet 2.2 while source 1 is still sending packet 1.2.

Since an agent starts to send whenever it has something to send, the delay is just the packet duration, i.e.,

$$d_{\mathrm{pALOHA},i} = \tau. \tag{6.14}$$

Originally, the collision probability of pure ALOHA was analyzed under the assumption of a Poisson arrival process.

---

[1]The *vulnerable period* is the time during which the sending of another agent leads to a collision.

Lemma 6.11 (Abramson (1970)). *For a Poisson arrival process with load ρ_Σ, the collision probability of pure ALOHA is*

$$q_{\text{pALOHA},i} = 1 - e^{-2\rho_\Sigma}. \tag{6.15}$$

Since the arrival process of event-based control is not a Poisson process, this lemma can not be applied directly. Nevertheless, Theorem 6.9 shows that the arrival process of event-based control converges to a Poisson process for $N \to \infty$. Consequently, if the number of agents is large, we can use (6.15) to approximate the collision probability. If this is not the case, or the exact collision probability is needed, we can use the following theorem from Sant (1980), where pure ALOHA with an arbitrary arrival process is studied.

Theorem 6.12 (Sant (1980)). *Let there be N users in the system and let each transmit packets of the same duration τ. For user j, let $f_j(x|\lambda_j)$ and $F_j(x|\lambda_j)$ denote the density and distribution functions, respectively, of the packet interarrival times. Let $1/\lambda_j$ be the average packet interarrival time for user j, so that he transmits at an average rate of λ_j. Then the steady-state probability of a packet collision for user i (assuming that he can possibly interfere with himself) is given by*

$$q_{\text{pALOHA},i} = 1 - \left(1 - \int_0^\tau f_i(x|\lambda_i)\mathrm{d}x\right)^2 \prod_{\substack{j=1,\\ j\neq i}}^{N} \left(1 - \lambda_j \int_0^{2\tau} [1 - F_j(x|\lambda_j)]\mathrm{d}x\right). \tag{6.16}$$

If user self-interference[2] is precluded, i.e., if for each j, $f_j(x|\lambda_j)$ and $F_j(x|\lambda_j)$ are zero for $x < \tau$, then the steady-state probability of a packet collision for user i reduces to

$$\tilde{q}_{\text{pALOHA},i} = 1 - \prod_{\substack{j=1,\\ j\neq i}}^{N} \left(1 - \lambda_j \int_0^{2\tau} [1 - F_j(x|\lambda_j)]\mathrm{d}x\right). \tag{6.17}$$

Note that there are two parameters that influence the collision probability: The packet duration τ and the sending rates λ_j. Thus, it is difficult to see how the collision probability scales with these parameters. Hence, we change the integration variables to reformulate Theorem 6.12 such that the collision probability depends only on the loads ρ_j.

Corollary 6.13. *In addition to the assumptions of Theorem 6.12, assume that for each user i, the distribution of the interarrival times belongs to a scale family with rate parameter λ_i. Then the steady-state probability of a packet collision for user i is*

$$q_{\text{pALOHA},i} = 1 - \left(1 - \int_0^{\rho_i} f_i(x|1)\mathrm{d}x\right)^2 \prod_{\substack{j=1,\\ j\neq i}}^{N} \left(1 - \int_0^{2\rho_j} [1 - F_j(x|1)]\mathrm{d}x\right). \tag{6.18}$$

---

[2] *Self-interference* means that a packet of an agent collides with another packet of the same agent.

Proof. This corollary follows by using the properties of a scale family and an appropriate change of the integration variables in Theorem 6.12. □

To get the collision probability from Theorem 6.12 and Corollary 6.14, we have to integrate the PDF and CDF of the interarrival times. Unfortunately, there exist distributions, where this can not be done analytically, e.g., the interevent time distribution of event-based traffic. In these cases, it is not possible to derive further analytical results. Moreover, Theorem 6.9 gives only an approximation for $N \to \infty$ and assumes that all agents send with the same rate. To get a simple description for a small number of agents, we next derive a lower and an upper bound for the collision probability of pure ALOHA. Although, these bounds are motivated by our analysis of event-based control with pure ALOHA, they hold for more general distributions of the interarrival times.

Lemma 6.14. *In addition to the assumptions of Theorem 6.12, assume that for each user i, the distribution of the interarrival times belongs to a scale family with rate parameter λ_i and $f_i(x|1) \leq 1$ holds for all i. Then the collision probability of pure ALOHA can be bounded by*

$$1 - \prod_{\substack{j=1,\\j\neq i}}^{N} a_{\mathrm{p}}(\rho_j) \leq q_{\mathrm{pALOHA},i} \leq 1 - \prod_{j}^{N} b_{\mathrm{p}}(\rho_j), \tag{6.19}$$

where

$$a_{\mathrm{p}}(\rho) := \begin{cases} 1 - 2\rho + 2\rho^2 & \text{for } \rho \leq 1/2 \\ 1/2 & \text{for } \rho > 1/2 \end{cases} \tag{6.20}$$

and

$$b_{\mathrm{p}}(\rho) := \begin{cases} 1 - 2\rho & \text{for } \rho \leq 1/2 \\ 0 & \text{for } \rho > 1/2. \end{cases} \tag{6.21}$$

The proof is given in Section D.6 of the appendix.

Remark 6.15. *The difference between the upper and lower bound is largest for $\rho_j \geq 1/2$ for all j. In this case, we have*

$$1 - (1/2)^{N-1} \leq q_{\mathrm{pALOHA},i} \leq 1. \tag{6.22}$$

Although these bounds are coarse for small N, they become tight for larger N. Moreover, for $N > 2$, the condition $\rho_j \geq 1/2$ means that the network is overloaded, which should be avoided. For realistic scenarios, where the network is not overloaded, i.e., $\rho_\Sigma \leq 1$, the bounds will be more tight than for the worst-case given by (6.22).

The previous results were motivated by our interest in event-based control over an ALOHA communication system but hold for much more general cases. For ease of reference, we specialize them to the case of event-based control in the following corollary.

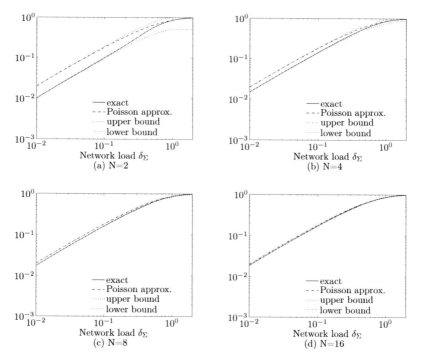

Figure 6.6.: Comparing the exact collision probability with the upper and lower bounds and the Poisson approximation for $N = 2, 4, 8$, and 16 agents.

Corollary 6.16. *Suppose, system* (6.1) *is controlled by an impulsive event-based control scheme with boundary increments* Δ_i. *The loop is closed over a shared communication system with a packet duration* τ *and pure ALOHA is used as MAC protocol. Then the steady-state probability of a packet collision for user i is*

$$q_{\text{pALOHA},i} = 1 - \left(1 - \int_0^{\rho_i} f_{\text{EB}}(x|1)\mathrm{d}x\right)^2 \prod_{\substack{j=1, \\ j\neq i}}^{N} \left(1 - \int_0^{2\rho_j}[1 - \int_0^x f_{\text{EB}}(t|1)\mathrm{d}t]\mathrm{d}x\right),$$

(6.23)

where $\rho_i = \frac{\tau\sigma_i^2}{\Delta_i^2}$ *is the load of agent i and f_{EB} given in* (6.7). *Moreover, these collision probabilities can be bounded as given in Lemma 6.14.*

Figure 6.6 shows the exact collision probability for event-based control with ALOHA, obtained from Corollary 6.16, the upper and lower bound of Lemma 6.14, and the

Poisson approximation from Lemma 6.11 for $N = 2$, 4, 8, and 16 agents. As already stated, the bounds are not tight for a small number of agents but become tighter as the number of agents increases. We also see that the Poisson approximation, which does not depend on the number of agents, becomes a good approximation for an already moderate number of agents.

Slotted ALOHA

The medium access of slotted ALOHA is very similar to pure ALOHA. In slotted ALOHA, the collision probability is reduced by restricting the time an user is allowed to send. Therefore, the time is divided into slots and an user is allowed to start sending only at the begin of a slot. Thereby, the vulnerable period is reduced from 2τ to τ. As a drawback, the delay is increased because the user has to wait for the begin of the next slot before sending, i.e., the delay is the packet duration plus the expected value for the end of the slot, which is half the packet duration.

Figure 6.5 shows an example. Here, packet 1.2 arrives during slot 4 and is thus sent in slot 5; packet 2.2 arrives during slot 5 and is thus sent in slot 6. Note that this avoids a collision when compared to pure ALOHA. Nevertheless, packet 1.3 and 2.3 arrive both during slot 7, are sent in slot 8, where they collide and are lost.

The expected delay of slotted ALOHA is

$$d_{\text{sALOHA},i} = 1.5\tau. \tag{6.24}$$

Lemma 6.17 (Roberts (1975)). *For a Poisson arrival process with load ρ_Σ, the collision probability of slotted ALOHA is*

$$q_{\text{sALOHA},i} = 1 - e^{-\rho_\Sigma}. \tag{6.25}$$

Another important difference to pure ALOHA are collisions due to self-interference. During our analysis of pure ALOHA, we assumed that all packets involved in a self-interference are lost. This should not be the case in slotted ALOHA, where a self-interference only occurs while the packet is waiting for the beginning of the next slot. This situation can be detected and handled such that the old packet is dropped and the new packet kept.

Theorem 6.18. *Let there be N users in the system and let each transmit packets with the same packet duration τ. Let the slot length be identical to the packet duration. For user j, let $f_j(x|\lambda_j)$ and $F_j(x|\lambda_j)$ denote the density and distribution functions, respectively, of the packet interarrival times. Let $1/\lambda_j$ be the average packet interarrival time for user j so that he transmits at an average rate of λ_j. If the newest packet survives a self-interference, then the steady-state probability of a packet collision for user i is given by*

$$q_{\text{sALOHA},i} = 1 - \frac{1}{\tau} \int_0^\tau \left(1 - \int_0^{\tau-x} f_i(t|\lambda_i)dt\right) dx \prod_{\substack{j=1, \\ j \neq i}}^N \left(1 - \lambda_j \int_0^\tau [1 - F_j(x|\lambda_j)]dx\right).$$

$$\tag{6.26}$$

If self-interference is precluded, then the steady-state probability of a packet collision for user i reduces to

$$\tilde{q}_{\text{sALOHA},i} = 1 - \prod_{\substack{j=1, \\ j \neq i}}^{N} \left(1 - \lambda_j \int_0^\tau [1 - F_j(x|\lambda_j)]\mathrm{d}x \right). \tag{6.27}$$

The proof is given in Section D.7 of the appendix.

Slotted ALOHA was developed to reduce the collision probability and it is a very well known fact that this is the case for Poisson traffic, as can be seen when comparing (6.15) and (6.25). The next lemma shows that this is also the case for general interarrival time distributions.

Lemma 6.19. *Suppose that $0 < f_i(x|\lambda_i) \ \forall x \neq 0$ and $0 < F_i(x|\lambda_i) \ \forall x \neq 0$. Then the collision probability of slotted ALOHA is smaller than the collision probability of pure ALOHA, i.e.,*

$$(a) \quad \tilde{q}_{\text{sALOHA},i} < \tilde{q}_{\text{pALOHA},i}, \qquad (b) \quad q_{\text{sALOHA},i} < q_{\text{pALOHA},i}.$$

Proof. First, note that $F_i(x|\lambda_i) < 1$ follows from $0 < f_i(x|\lambda_i)$. Now, part (a) follows from the fact that $0 < F_j(x|\lambda_j) < 1 \ \forall x \neq 0$, and therefore $\int_0^\tau [1 - F_j(x|\lambda_j)]\mathrm{d}x < \int_0^{2\tau} [1 - F_j(x|\lambda_j)]\mathrm{d}x$.

To prove Part (b), we have to compare the collision probabilities due to self-interference. Since $\int_0^{\tau-x} f_i(t|\lambda_i)\mathrm{d}t < \int_0^\tau f_i(t|\lambda_i)\mathrm{d}t \ \forall x < \tau$, we have $\frac{1}{\tau} \int_0^\tau 1 - \int_0^{\tau-x} f_i(t|\lambda_i) \leq \frac{1}{\tau} \int_0^\tau 1 - \int_0^\tau f_i(t|\lambda_i)\mathrm{d}t\mathrm{d}x = 1 - \int_0^\tau f_i(t|\lambda_i)\mathrm{d}t$. $\qquad \square$

Again, we can give the collision probability in terms of the load and also a lower and upper bound on the collision probability.

Corollary 6.20. *In addition to the assumptions of Theorem 6.18, assume that for each user i, the distribution of the interarrival times belongs to a scale family with rate parameter λ_i. Then the collision probability for user i is*

$$q_{\text{sALOHA},i} = 1 - \frac{1}{\rho_i} \int_0^{\rho_i} \left(1 - \int_0^{\rho_i - x} f_i(t|1)\mathrm{d}t \right) \mathrm{d}x \prod_{\substack{j=1, \\ j \neq i}}^{N} \left(1 - \int_0^{\rho_j} [1 - F_j(x|1)]\mathrm{d}x \right).$$

$$\tag{6.28}$$

Proof. This corollary follows by using the properties of a scale family and an appropriate change of the integration variables in Theorem 6.18. $\qquad \square$

Lemma 6.21. *In addition to the assumptions of Theorem 6.18, assume that for each user i, the distribution of the interarrival times belongs to a scale family with rate* *parameter λ_i and $f_i(x|1) \leq 1$ holds for all i. Then the collision probability of slotted ALOHA can be bounded by*

$$1 - \prod_{\substack{j=1, \\ j \neq i}}^{N} a_s(\rho_j) \leq q_{sALOHA,i} \leq 1 - \prod_{j}^{N} b_s(\rho_j), \tag{6.29}$$

where

$$a_s(\rho) := \begin{cases} 1 - \rho + \frac{1}{2}\rho^2 & \text{for } \rho \leq 1 \\ \frac{1}{2} & \text{for } \rho > 1 \end{cases} \tag{6.30}$$

and

$$b_s(\rho) := \begin{cases} 1 - \rho & \text{for } \rho \leq 1 \\ 0 & \text{for } \rho > 1. \end{cases} \tag{6.31}$$

The proof is given in Section D.8 of the appendix.

Again, the previous results were motivated by our interest in event-based control over a slotted ALOHA communication system but hold for much more general cases. For ease of reference, we specialize them to the case of event-based control in the following corollary.

Corollary 6.22. *Suppose, system* (6.1) *is controlled by an impulsive event-based control scheme with boundary increments Δ_i. The loop is closed over a shared communication system with a packet duration τ and slotted ALOHA is used as MAC protocol. Then the steady-state probability of a packet collision for user i is*

$$q_{sALOHA,i} = 1 - \frac{1}{\rho_i} \int_0^{\rho_i} \left(1 - \int_0^{\rho_i - x} f_{EB}(t|1)dt \right) dx \prod_{\substack{j=1, \\ j \neq i}}^{N} \left(1 - \int_0^{\rho_j} [1 - \int_0^{x} f_{EB}(t|1)dt]dx \right). \tag{6.32}$$

where $\rho_i = \frac{\tau \sigma_i^2}{\Delta_i^2}$ is the load of agent i and f_{EB} given in (6.7). *Moreover, these collision probabilities can be bounded as given in Lemma 6.21.*

Queueing System

To resolve conflicts in a contention based communication system, there is a tradeoff between loss and delay. Within the previous section on slotted ALOHA, we already saw that the loss probability can be reduced by introducing a small delay. In this section, the communication system is modeled as a FIFO queue, i.e., the packets are sent over the shared medium in the same order they are generated. In doing so, the packets are delayed but never lost. Consequently, this model can be interpreted as the opposite of pure ALOHA, where packets are lost but not delayed due to the medium access.

From a more practical perspective, this model is motivated by todays Ethernet, which is no longer formed by connecting all users to a shared medium and thereby forming a global collision domain. Instead, all users are connected to a switch, which resolves arbitration conflicts by buffering the packets in a queue. Thus, todays Ethernet can be modeled as a queueing system.

Figure 6.5 shows an example. When packet 1.2 arrives, the medium is idle and the packet is immediately sent. When packet 2.2 arrives, the medium is found busy and thus the transmission is delayed until packet 1.2 is finished.

For analyzing the queueing system, we assume that the arrival process is Poisson (Markovian). Moreover, Assumption 6.1 states that the packet duration is constant, i.e., the service process is deterministic. Finally, since there is one shared medium, we have one service unit. Thus, the considered queueing system is an M/D/1 queue.

Lemma 6.23 (Kleinrock (1975)). *For an infinite M/D/1 queue, the delay and loss are*

$$d_{\text{iQueue},i} = \begin{cases} \frac{2-\rho_\Sigma}{2(1-\rho_\Sigma)}\tau & \text{for } \rho_\Sigma < 1 \\ \infty & \text{for } \rho_\Sigma \geq 1 \end{cases} \tag{6.33}$$

$$q_{\text{iQueue},i} = 0. \tag{6.34}$$

Since an infinite queue is not realistic and gives an infinite delay when the network is overloaded, we now look at a finite queue. Although the results for the infinite M/D/1 queue are relatively old and well known, the finite M/D/1 queue has been solved only recently in Brun and Garcia (2000); Davis and Howl (1997).

Lemma 6.24 (Brun and Garcia (2000)). *The expected delay and loss probability for a finite M/D/1 queue of size M (waiting room + processing unit) are*

$$d_{\text{fQueue},i} = \left(M - \frac{\sum_{k=0}^{M-1} b_k - M}{\rho_\Sigma b_{M-1}} \right) \tau, \tag{6.35}$$

$$q_{\text{fQueue},i} = 1 - \frac{b_{M-1}}{1 + \rho_\Sigma b_{M-1}}, \tag{6.36}$$

where the coefficients b_n are

$$b_n = \sum_{k=0}^{n} \frac{(-1)^k}{k!}(n-k)^k e^{(n-k)\rho_\Sigma} \rho_\Sigma^k. \tag{6.37}$$

By varying the queue size, it is possible to trade loss against delay, see also Sommer and Blind (2007).

Erlang's Loss Model

Our motivation to model the medium access with Erlang's loss model is twofold. First, Erlang's loss model is indeed a queue with a size of one, i.e., only the processing unit, which represents the shared medium. However, our main motivation is to model a variation of non-persistent CSMA. In the original non-persistent CSMA model, the packet is sent when the medium is found idle and retransmitted after a random time whenever the medium is found busy, see, e.g., Kleinrock and Tobagi (1975); Rom and Sidi (1990). However, as stated in Theorem 6.10, for our setup, it is better to drop a packet and let the underlying event-based control generate the next packet than retransmitting it. Thus, we propose to use a variation of non-persistent CSMA, where the packet is dropped when the medium is found busy. This is indeed Erlang's loss model with one service unit, as first presented in Erlang (1917).

Note the difference to the ALOHA protocol: In ALOHA all packets involved in a collision are lost. This means that a packet can be destroyed by an agent that starts to send at some later time. In contrast, in Erlang's loss model, a packet is only sent when the medium is idle. Thus, all packets that get access to the medium will be transmitted successfully.

Figure 6.5 shows an example. When packet 1.2 arrives, the medium is idle and the packet is immediately sent. When packet 2.2 arrives, the medium is busy and packet 2.2 is dropped. Note that in contrast to the ALOHA protocol, packet 1.2 is not affected by the arrival of packet 2.2.

Lemma 6.25 (Erlang (1917)). *The expected delay and loss probability for Erlang's loss model with a Poisson arrival and deterministic service unit is*

$$d_{\text{Erlang},i} = \tau \tag{6.38}$$

$$q_{\text{Erlang},i} = \frac{\rho_\Sigma}{1 + \rho_\Sigma}. \tag{6.39}$$

6.5. Control and Communication

In this section, we use the results from the previous two sections to derive the optimal load and the minimal cost of time-triggered and event-based control with different communication systems.

When designing a networked control system, both, the controller and the communication system must be parameterized. The speed of the communication system is given by the packet duration τ. For the controller, the sampling time or boundary increment, which directly gives the expected interevent time, must be chosen. Since both parameters affect the cost of the closed loop system directly or indirectly, the packet duration τ and the sampling time $T_{\text{TT},i}$ or expected interevent time $T_{\text{EB},i}$ must be considered together. This can be done by using their ratio, i.e., the load ρ_i, as parameter. As already stated in Remark 6.6, for event-based control, an arbitrary value of the load ρ_i can be achieved by the choice of the boundary increment Δ_i. Obviously, for time-triggered control, the sampling time $T_{\text{TT},i}$, and thereby the load, can be chosen directly.

For comparing the different control and communication schemes, we thus use the normalized cost, which we define as follows.

Definition 6.26. *The* normalized cost *is*

$$\tilde{J}_i := \frac{J_i}{\tau \sigma_i^2}.$$

The normalized cost of time-triggered and event-based control becomes

$$\tilde{J}_{\mathrm{TT},i} := \frac{J_{\mathrm{TT},i}}{\tau \sigma_i^2} = \frac{1}{2\rho_i} + \frac{q_i}{\rho_i(1-q_i)} + \frac{d_i}{\tau}, \tag{6.40}$$

$$\tilde{J}_{\mathrm{EB},i} := \frac{J_{\mathrm{EB},i}}{\tau \sigma_i^2} = \frac{1}{6\rho_i} + \frac{q_i}{\rho_i(1-q_i)} + \frac{d_i}{\tau}. \tag{6.41}$$

Since the problem setup contains N systems, the normalized cost of all agents together, i.e., $\tilde{J}_\Sigma = \sum_{i=1}^{N} \tilde{J}_i$ will be used to compare the different setups.

Note that for all communication protocols, that were presented in the previous section, the delay is a multiple of the packet duration and the loss probability depends only on the load. When taking this into account, we see that the normalized cost depends only on the load. Thus, we use the load as tuning parameter to minimize the normalized cost.

6.5.1. Time-Triggered Control with a Deterministic MAC

Time-Triggered Control with TDMA

Note that the time-triggered sampling scheme determines the time between the samples of an individual agent but not the time between the samples of the different systems. When taking the medium access into account, the relative sampling time between the different systems must also be considered. The two most common approaches are a *synchronous* and a *round-robin* sampling scheme. In the synchronous scheme, all systems are sampled at the same time, as depicted in Figure 6.7a. Note that this scheme introduces some delay between the sampling and the transmission of the corresponding packet, especially when the packet is transmitted at the end of a period. In the round-robin scheme, the system is sampled immediately before the begin of the corresponding slot, i.e., the sampling and the end of the transmission of the previous agent coincide, see Figure 6.7b. In doing so, the delay is just the transmission time of a packet.

Theorem 6.27. *Suppose, there are N agents, each uses the time-triggered control scheme to control system* (6.1). *Moreover, these agents are sampled with the round-robin scheme and TDMA is used to manage access to the shared medium, then:*

(i) The normalized cost for agent i is

$$\tilde{J}_{\mathrm{TDMA},i} = \frac{1}{2\rho_i} + 1. \tag{6.42}$$

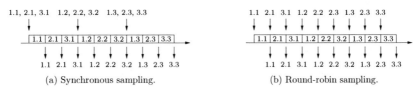

(a) Synchronous sampling.　　　　　　(b) Round-robin sampling.

Figure 6.7.: Time-triggered sampling schemes.

(ii) The cost of all agents together is minimal for $\rho_i = \rho_j$.

(iii) The optimal network load is

$$\rho_{\text{TDMA},\Sigma}^* = 1. \tag{6.43}$$

(iv) The minimal normalized cost for each agent is

$$\tilde{J}_{\text{TDMA},i}^* = \frac{1}{2}N + 1. \tag{6.44}$$

Proof. (i) follows by using (6.9) and (6.8) in (6.4). The rest of this theorem follows from the Lagrange dual function of (6.42) with the constraint (6.10). □

Note that the optimal load of each agent is $\rho_{\text{TDMA},i}^* = 1/N$.

Time-Triggered Control with FDMA

The analysis of time-triggered control with FDMA is similar to the analysis of time-triggered control with TDMA. The only difference is the longer packet transmission time, which is $m_i\tau$ for FDMA instead of τ for TDMA.

Theorem 6.28. *Suppose, there are N agents, each uses the time-triggered control scheme to control system (6.1). Moreover, FDMA is used to manage access to the shared medium, then:*

(i) The normalized cost for agent i is

$$\tilde{J}_{\text{FDMA},i} = \frac{1}{2\rho_i} + m_i. \tag{6.45}$$

(ii) The cost of all agents together is minimal for $\rho_i = \rho_j$.

(iii) The optimal network load is

$$\rho_\Sigma^* = 1. \tag{6.46}$$

(a) Low load.   (b) High load.

Figure 6.8.: An example of event-based control with TDMA for agent 1.

(iv) The minimal normalized cost for each agent is

$$\tilde{J}^*_{\text{FDMA},i} = \frac{3}{2}N. \tag{6.47}$$

Proof. (i) follows by using (6.12) and (6.11) in (6.4). The rest of this theorem follows from the Lagrange dual function of (6.45) with the constraint (6.13). □

Again, the optimal load of each agent is $\rho^*_{\text{FDMA},i} = 1/N$.

6.5.2. Time-Triggered Control with a Contention Based MAC

In general, it is possible to use a contention based MAC to transmit the packets generated by a time-triggered sampling scheme. However, when done properly, the resulting medium access turns out to be identical to time-triggered control with TDMA as explained in the following. When the systems are sampled synchronously, all agents send at the same time. When the arbitration conflicts are not resolved, as would be the case with pure ALOHA, all packets are lost due to Assumption 6.2. Obviously, this approach makes no sense. However, when a global queue is used to resolve the arbitration conflicts, as is the case for a modern Ethernet like MAC layer, the resulting transmission times will be similar to TDMA but the packets will be delayed. When the different systems are sampled round-robin, the resulting transmission times are identical to TDMA and the delay is just the packet duration. Thus, when properly designed, time-triggered control with a contention based MAC will be similar to time-triggered control with TDMA.

6.5.3. Event-Based Control with a Deterministic MAC

In general, it is possible to use a deterministic MAC protocol to transmit the packets generated by an event-based sampling scheme. However, it is more efficient to use time-triggered control with a deterministic MAC than event-based control with a deterministic MAC, as explained in the following for TDMA. First, suppose that the offered load is low. An example of this case is depicted in Figure 6.8a. Note that not all slots are used and there is a relatively long delay between a transmission request and its start. In this case, time-triggered sampling would result in a better performance. Now, suppose that the offered load is increased and arbitration conflicts are resolved such that the last packet before the reserved time slot is sent. The case of a relatively high load is depicted in Figure 6.8b, where the dashed arrows indicate dropped requests and the solid arrows indicate transmitted requests. In this example, all reserved slots are used but there is still a delay between the transmission request and its start. It is easy to see that all slots are used and the delay between the transmission request and its start approaches zero when the offered load approaches infinity. When looking only at the successfully transmitted packets, this is similar to time-triggered control with TDMA.

To sum up, event-based control with TDMA becomes similar to time-triggered control with TDMA when the offered load becomes infinite. However, as long as the offered load is finite, time-triggered control with TDMA gives a better performance than event-based control with TDMA.

6.5.4. Event-Based Control with a Contention Based MAC

Obviously, it is difficult to derive analytically exact results for event-based control with a contention based communication system. However, Theorem 6.9 states that the arrival process of all agents together converges to a Poisson process for $N \to \infty$. To derive the optimal load and minimal cost for a finite number of agents, we thus approximate the arrival process of event-based control by a Poisson process, even when the number of agents is finite.

Moreover, we assume that packet loss and delay are independent of the state of the system. Note that this is not always the case as can be seen from the example of one agent with an infinite queue. When the state reaches the k-th bound, we know that there are at least $k - 1$ packets missing, which must be in the queue. In this case, the expected delay depends on the state of the system. However, we claim that this effect can be neglected for practical applications when the number of agents is large and the offered load of each agent is small.

Event-Based Control with Pure ALOHA

Theorem 6.29. *Suppose, there are N agents, each uses the event-based control scheme to control system (6.1) and the arrival process of all agents is approximated by a Poisson process. Moreover, pure ALOHA is used to manage access to the shared medium, then:*

(i) The normalized cost for agent i is

$$\tilde{J}_{\text{pALOHA},i} = \frac{1}{6\rho_i} + \frac{1 - e^{-2\rho_\Sigma}}{\rho_i e^{-2\rho_\Sigma}} + 1 = \frac{6e^{2\rho_\Sigma} - 5}{6\rho_i} + 1. \qquad (6.48)$$

(ii) The cost of all agents together is minimal for $\rho_i = \rho_j$.

(iii) The optimal network load fulfils

$$5 + 6(2\rho^*_{\text{pALOHA},\Sigma} - 1)e^{2\rho^*_{\text{pALOHA},\Sigma}} = 0. \tag{6.49}$$

Numerically, this is $\rho^*_{\text{pALOHA},\Sigma} \approx 0.2445$.

(iv) The optimal loss probability fulfills

$$5q^*_{\text{pALOHA},i} = 6\ln(1 - q^*_{\text{pALOHA},i}) + 1 = 0. \tag{6.50}$$

Numerically, this is $q^*_{\text{pALOHA},i} \approx 0.3867$.

(v) The minimal normalized cost for each agent is

$$\tilde{J}^*_{\text{pALOHA},i} = 2e^{2\rho^*_{\text{pALOHA},\Sigma}}N + 1 \approx 3.2612N + 1. \tag{6.51}$$

Proof. (i) follows by using (6.14) and (6.15) in (6.41). (ii) follows from the Lagrange dual function of (6.48) with the constraint $\rho_\Sigma = \sum \rho_i$. Now, (iii) follows by using $\rho_\Sigma = N\rho_i$ and checking the first and second derivative of (6.48). Finally, (iv) follows by using (6.15) in (6.49) and (v) follows by using (6.15) in (6.48). □

As can be seen from (6.49), the optimal network load is independent of the number of agents. As a consequence thereof, the optimal loss probability is also independent of the number of agents, as can be seen in (6.50). However, the optimal load of each agent, i.e., $\rho^*_{\text{pALOHA},\Sigma}/N$ depends on the number of agents. Consequently, an agent must know the number of agents to send with the optimal load.

Another interesting observation is the fact that the optimal network load is less than half the network load that gives the highest throughput, which is $\rho_\Sigma = 0.5$, see, e.g., Abramson (1970). This indicates that in the considered setup reliability is more important than mere throughput.

Event-Based Control with Slotted ALOHA

Theorem 6.30. *Suppose, there are N agents, each uses the event-based control scheme to control system (6.1) and the arrival process of all agents is approximated by a Poisson process. Moreover, slotted ALOHA is used to manage access to the shared medium, then:*

(i) The normalized cost of agent i is

$$\tilde{J}_{\text{sALOHA},i} = \frac{1}{6\rho_i} + \frac{1 - e^{-\rho_\Sigma}}{\rho_i e^{-\rho_\Sigma}} + 1.5 = \frac{6e^{\rho_\Sigma} - 5}{6\rho_i} + 1.5. \tag{6.52}$$

(ii) The cost of all agents together is minimal for $\rho_i = \rho_j$.

(iii) *The optimal network load fulfils*

$$5 + 6(\rho^*_{\text{sALOHA},\Sigma} - 1)e^{\rho^*_{\text{sALOHA},\Sigma}} = 0. \tag{6.53}$$

Numerically, this is $\rho^*_{\text{sALOHA},\Sigma} \approx 0.4889$.

(iv) *The optimal loss probability fulfills*

$$5q^*_{\text{sALOHA},i} = 6\ln(1 - q^*_{\text{sALOHA},i}) + 1 = 0. \tag{6.54}$$

Numerically, this is $q^*_{\text{sALOHA},i} \approx 0.3867$.

(v) *The minimal normalized cost for each agent is*

$$\tilde{J}^*_{\text{sALOHA},i} = e^{\rho^*_{\text{sALOHA},\Sigma}} N + 1.5 \approx 1.6306N + 1.5. \tag{6.55}$$

The proof is similar to the one of Theorem 6.29 and thus omitted.

Comparing Theorem 6.29 and Theorem 6.30, we see that the optimal load for event-based control with slotted ALOHA is twice the optimal load for event-based control with pure ALOHA. More surprisingly, the optimal loss probability of event-based control with pure ALOHA and the optimal loss probability of event-based control with slotted ALOHA are identical.

Event-Based Control with an Infinite Queue

Theorem 6.31. *Suppose, there are N agents, each uses the event-based control scheme to control system (6.1) and the arrival process of all agents is approximated by a Poisson process. Moreover, an infinite queue is used to model the medium access, then:*

(i) *The normalized cost of agent i is*

$$\tilde{J}_{\text{iQueue},i} = \frac{1}{6\rho_i} + \frac{2 - \rho_\Sigma}{2(1 - \rho_\Sigma)}. \tag{6.56}$$

(ii) *The cost of all agents together is minimal for $\rho_i = \rho_j$.*

(iii) *The optimal network load is*

$$\rho^*_{\text{iQueue},\Sigma} = \begin{cases} \frac{1}{2} & \text{for } N = 3 \\ \frac{N - \sqrt{3N}}{N - 3} & \text{for } N \neq 3. \end{cases} \tag{6.57}$$

(iv) *The optimal delay is*

$$d^*_{\text{iQueue},i} = \left(\frac{N + \sqrt{3N} - 6}{2(-3 + \sqrt{3N})} \right) \tau. \tag{6.58}$$

(v) The minimal normalized cost for each agent is

$$\tilde{J}^*_{\text{iQueue},i} = \frac{N-3}{6(1-\sqrt{3/N})} + \frac{\sqrt{N}+\sqrt{3}-6/\sqrt{N}}{2(-3/\sqrt{N}+\sqrt{3})} \approx \frac{N}{6} + \frac{\sqrt{N}}{2\sqrt{3}} \quad \textit{for large } N.$$

(6.59)

Proof. We use (6.33) in (6.41) to get (i). (ii) follows from the Lagrange dual function of (6.56) and the constraint $\rho_\Sigma = \sum \rho_i$. Now, (iii) follows by checking the first and second derivative of (6.56). Finally, (iv) follows by using (6.57) in (6.56) and (v) follows by using (6.57) and (6.58) in (6.56). □

Event-Based Control with a Finite Queue

For a finite queue, it is difficult to get exact analytical results based on Lemma 6.24. Anyway, the most interesting cases are a lowly and highly loaded queue. Hence, we restrict our discussion to these cases. Thereby, we assume that all agents send with the same rate.

If the load is not too large, then the performance of event-based control with a finite queue is similar to the performance of event-based control with an infinite queue. Obviously, the difference becomes smaller as the size of the queue increases. On the other hand, if the queue is heavily overloaded, i.e., $\rho_i \to \infty$, most packets find a full queue and are lost. Whenever a packet leaves the queueing system, the next arriving packet will be enqueued and finds $M-1$ other packets in the queue. Thus, the delay of this packet will be $M\tau$. Consequently, the additional relative cost due to delay becomes

$$\tilde{J}_{\text{fQueue,delay},i} \to M \qquad \text{for } \rho_i \to \infty.$$

(6.60)

Moreover, for $\rho_i \to \infty$, the loss rate is such that the load of the non-lost packets becomes one, i.e.,

$$\rho_\Sigma(1-q_i) = 1 \quad \Rightarrow \quad \rho_i(1-q_i) = 1/N.$$

(6.61)

Consequently, the additional relative cost due to packet losses are

$$\tilde{J}_{\text{fQueue,loss},i} = \frac{q_i}{\rho_i(1-q_i)} \to N \qquad \text{for } \rho_i \to \infty.$$

(6.62)

Adding (6.60) and (6.62), we get for the normalized cost

$$\tilde{J}_{\text{fQueue},i} \to N + M \qquad \text{for } \rho_i \to \infty.$$

(6.63)

The same result can also be obtained by using Lemma 6.24. Therefore, it is crucial to note that $b_l < b_k$ for $l < k$ and that b_n is increasing with increasing ρ_i. Thus, we have

$$\tilde{J}_{\text{fQueue,loss},i} = \frac{q_i}{\rho_i(1-q_i)} = \frac{1+\rho_i N b_{M-1} - b_{M-1}}{\rho_i b_{M-1}} \to N \qquad \text{for } \rho_i \to \infty.$$

Moreover, observe that

$$\frac{\sum_{k=0}^{M-1} b_k - M}{N\rho_i b_{M-1}} \to 1 \qquad \text{for } \rho_i \to \infty$$

and thus

$$\tilde{J}_{\text{fQueue,delay},i} \to M \qquad \text{for } \rho_i \to \infty.$$

To sum up, if the finite queue is properly sized and the load is low, then the cost of event-based control with a finite queue is similar to the cost of event-based control with an infinite queue. If the finite queue is heavily overloaded, i.e., $\rho_i \to \infty$, then the cost converges to the sum of the number of agents and the queue size, i.e., the cost remains finite.

Event-Based Control with Erlang's Loss Model

Theorem 6.32. *Suppose, there are N agents, each uses the event-based control scheme to control system (6.1) and the arrival process of all agents is approximated by a Poisson process. Moreover, Erlang's loss model is used to model the medium access, then:*

(i) The normalized cost of agent i is

$$\tilde{J}_{\text{Erlang},i} = \frac{1}{6\rho_i} + \frac{\rho_\Sigma}{\rho_i} + 1. \tag{6.64}$$

(ii) The cost of all agents together is minimal for $\rho_i = \rho_j$.

(iii) The minimal normalized cost is achieved for

$$\rho_i \to \infty. \tag{6.65}$$

(iv) The minimal normalized cost is

$$\tilde{J}_{\text{Erlang},i}^* = N + 1 \tag{6.66}$$

Proof. (i) follows by using (6.38) and (6.39) in (6.41). (ii) follows from the Lagrange dual function of (6.64) with the constraint $\rho_\Sigma = \sum \rho_i$. Finally, (iii) and (iv) follow directly from (6.64). $\qquad \square$

As expected, the normalized cost for $\rho_i \to \infty$ as given in (6.66) is equal to the normalized cost obtained from a finite queue with size $M = 1$ for $\rho_i \to \infty$, as given in (6.63).

For event-based control $\rho_i \to \infty$ requires $\Delta_i \to 0$, i.e., the boundary increment Δ_i is so small that the agent tries to send all the time. Due to the contention based medium access, the choice of the sending agent will be random. This can be interpreted as a third sampling scheme: random sampling.

Table 6.1.: The minimal normalized cost of the different schemes.

control scheme	communication scheme	minimal normalized cost $\bar{J}_i^*$
time-triggered	TDMA	$0.5N + 1$
	FDMA	$1.5N$
event-based	pure ALOHA	$2e^{2\rho_{\mathrm{pALOHA},\Sigma}^*}N + 1 \approx 3.26N + 1$
	slotted ALOHA	$e^{\rho_{\mathrm{sALOHA},\Sigma}^*}N + 1.5 \approx 1.63N + 1.5$
	infinite queue	$\dfrac{N-3}{6(1-\sqrt{3/N})} + \dfrac{\sqrt{N}+\sqrt{3}-6/\sqrt{N}}{2(-3/\sqrt{N}+\sqrt{3})}$
		$\approx \frac{1}{6}N + \frac{\sqrt{N}}{2\sqrt{3}}$ for large N
	Erlang's loss model	$N + 1$

6.5.5. Comparison of the Different Control and Communication Schemes

Table 6.1 summarizes the minimal cost of the different control and communication schemes. For all considered control and communication schemes, except event-based control with a queueing system, the minimal cost is affine in the number of agents and thus easy to compare.

Figure 6.9 shows the normalized cost over the network load for $N = 10$ agents. When the network load is relatively low, the additional costs due to loss and delay are not significant. Thus, the cost is mostly determined by the sampling strategy. The influence of the communication system becomes significant for higher network loads, where the additional costs due to loss and/or delay are large. This is best seen when comparing event-based and time-triggered control. Although event-based control gives a better performance than time-triggered control for a low network load, it becomes worse than time-triggered control when the network load is high.

For event-based control with ALOHA, the loss probability is relatively high and thus also the additional cost due to loss. As a consequence thereof, the minimal cost of event-based control with ALOHA is larger than the minimal cost of time-triggered control with TDMA or FDMA.

When using event-based control with a queueing system, the minimal cost is smaller than the minimal cost achievable by time-triggered control with TDMA or FDMA. However, when the network capacity is exceeded, i.e., $\rho_\Sigma \geq 1$, the cost will be infinite due to an infinite delay. This problem does not occur when a finite queue is used, where the cost remains finite, even for $\rho_i \to \infty$, see Section 6.5.4. The longer the queue, the more similar it is to an infinite queue. The minimal achievable cost will be smaller for a longer queue but the cost will be larger when the communication system is overloaded. Consequently, the dimensioning of the queue is not a trivial task.

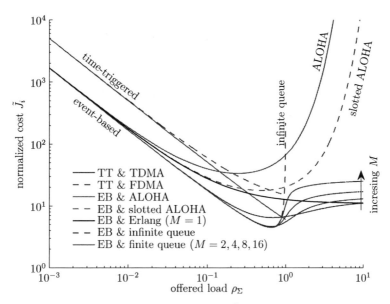

Figure 6.9.: The normalized cost $\tilde{J}_i$ for $N = 10$ agents.

6.6. Summary

In this chapter, we analyzed the effects of the communication system and the inter-action between control and communication for networked control systems. Therefore, we fixed the control problem and considered different communication systems. While comparing time-triggered and event-based control, we observed that the sampling strategy affects the performance not only directly but also indirectly via the loss and delay of the communication system. Thereby, not only the control strategy is impor-tant but also the choice of the communication system because it determines how loss and delay depend on the network load.

We also discussed how the performance depends on the load. As long as the network load is low, packets are rarely lost due to collisions and the delay due to the medium access is short. In this case, the performance is almost solely determined by the sampling strategy and the sending rate. Moreover, for a given loss probability and delay, the cost is decreasing with the load. Thus, it seems to be obvious to increase the load to improve the performance. Interestingly, this is no longer the case when the medium access is taken into account. For deterministic MAC protocols there is a hard limit for the load because it is not possible to overload the communication system. For contention based MAC protocols, we showed that loss and delay increase with the load and become significant when the network load is high. Except for event-based control with Erlang's loss model, the optimal network load is finite. The cost increases with the load when the optimal load is exceeded. Thus, it is not worth to increase the load beyond this point. Nevertheless, this is not the case for event-based control with Erlang's loss model. Here, the performance is optimal for an infinite network load, which corresponds to a random sampling.

Chapter 7.

Conclusions

7.1. Summary

We presented several approaches to optimize the communication system to improve the control performance. We started with the optimization within the presentation layer in Chapter 2, where we presented an approach to improve remote estimation over a communication system with packet loss. We showed that it is possible to get better estimates, or to make the Kalman filter more robust against packet loss, by preprocessing the measurements before sending them. Compared to other approaches, that also aim to improve remote estimation over a communication system with packet loss, the preprocessing, which must be performed by the sensor, was relatively simple; it was just a linear transformation.

Another approach to improve remote estimation is the usage of a smart sensor that calculates the state estimate within the sensor. Obviously, this requires some computational power within the sensor. In recent literature, it was shown that optimal estimates at the remote estimator can be achieved by a smart sensor that calculates the state estimate and sends it to the remote estimator or by a sensor that always sends all measurements. In Chapter 3, we showed that the same optimal estimates can be achieved without preprocessing the measurements or huge data transfers by simply acknowledging successfully received measurements and adding all non-acknowledged measurements to the current measurement packet.

In Chapter 4, we again used the idea of acknowledging and automatically retransmitting lost packets, but also considered sending each packet multiple times. However, instead of interpreting this as delay, we increased the sampling time to allow a limited number of transmissions within each sampling interval. In doing so, retransmitted packets arrive on time and the sampling time is determined by the transmission times of the communication system and the number of transmissions per sampling interval. Since increasing the sampling time now improves the reliability of the communication system but generally degrades the control performance, the optimal choice of the sampling time becomes an interesting problem. Thus, we introduced and analyzed four different configurations of the transport layer. Finally, with the help of three examples, we showed that the control performance can be improved by a proper design of the transport layer, even while reducing the expected packet rate.

Based on the idea that loss and delay follow from the design of the communication system, we considered the optimal routing through the communication system in Chapter 5. Here, loss and delay of an end-to-end connection can be determined from the loss and delay of the traversed links. To analyze this problem mathematically, we modeled the underlying communication system as a graph and were then able to formulate the routing and controller design problem as an optimization problem.

When looking at the details of a communication system, it becomes clear that packets are either lost due to random noise or lost and/or delayed due to arbitration conflicts and approaches to resolve them. Consequently, packet loss and delay depend on the design of the communication system and its usage. When this is taken into account, the interaction between control and communication becomes a very interesting and relevant problem when designing a networked control system. The control engineer has to design the communication system, and thereby chooses the arbitration protocol, but is also responsible for the usage of the communication system. Thus, we considered the design of the MAC layer in Chapter 6 by analyzing and comparing time-triggered and event-based control with a shared communication system. We showed that the difference in the traffic pattern of time-triggered and event-based control significantly affects the choice of the MAC protocol. Thus, time-triggered control is best suited for a deterministic MAC protocol and event-based control should be used with a contention based MAC protocol. For both sampling strategies, the performance increases with the load but decreases with loss and delay. Since loss and delay are independent of the load when a deterministic MAC protocol is used, time-triggered control with a deterministic MAC protocol is optimal when the communication system is fully utilized. On the other hand, when using a contention based MAC protocol, loss and/or delay increase with the load. Thus, for event-based control with a contention based MAC protocol, there is an optimal load, which obviously depends on the MAC protocol. We showed that the best performance is achieved by an event-based controller with a global queue.

Although the presented approaches are located within different layers of the ISO/OSI reference model, they are all based on relatively well known ideas in the field of communication theory, e.g., the method of acknowledging successfully transmitted packets and retransmitting lost packets in Chapter 3 and 4. Thus, we conclude that for further improvements in the field of networked control systems, it will be crucial to give up the sharp separation between control and communication and use methods from the field of communication theory to improve the control performance.

7.2. Outlook

Throughout this thesis, packet loss was modeled by a Bernoulli process, i.e., packet loss was assumed to be independent and identically distributed (iid). Moreover, packet delay was modeled by a constant delay. Obviously, when looking at real communication systems, this is often not the case. First, note that the Quality of Service (QoS) of the communication system is in general not constant due to a varying load of the communication system. This could be taken into account by using time varying arrival probabilities and delays. Furthermore, the loss probability and delay of two consecutive packets will be correlated when they take the same route and wait in the same queue. Moreover, packets might be scheduled such that they get a different priority after each loss. Such dependencies between consecutive packets can be taken into account by modelling the communication system with the help of a Markov chain. In doing so, the entire system becomes a Markov Jump Linear System (MJLS) and can be studied within this framework. Finally, there are also dependencies in space. E.g., packets that take the same route could be transmitted within one larger super-packet. In doing so, either all packets arrive or all are lost. On the other extreme, when the bandwidth is very limited such that only one packet can be transmitted at each time, the packets must be scheduled accordingly. In this case, when one packet arrives, we know for sure that the others are lost. Such dependencies in space can be captured by the model of the communication system of Chapter 2 and a proper choice for the probabilities that a particular erasure matrix is chosen.

In Chapter 4, we assumed that the state can be measured directly. Obviously, this might not be realistic and should be extended to the case of noisy measurements. Moreover, the model of the underlying communication system was relatively simple. We assumed that the transmission times as well as the packet arrival probabilities were constant and known. In reality, transmission times are not constant and might be better described by a probability distribution. In doing so, the loss probability follows by the choice of the maximal waiting time, i.e., the time the controller waits until it declares a packet as lost. Taking this into account, the choice of the maximal waiting time would be a very interesting problem.

In Chapter 5, we formulated the controller design, its placement within the underlay communication system, and the routing between sensor, controller, and actuator as an optimization problem. Unfortunately, the resulting optimization problem is difficult to solve due to its nonlinear cost function and its integer constraints. For practical applications, it will be necessary to find methods to solve this optimization problem more efficiently or find good approximations of the optimal solution within a reasonable time. Moreover, we considered a relatively simple scenario in Chapter 5: There was only one control system with one sensor and one actuator, the capacities of the underlay links were infinite, and the loss and delay of a link independent of its usage. Studying more realistic scenarios, where several loops are closed over the same communication system, and more detailed link models, e.g., links with a finite capacity or links where the loss and delay depend on their usage, will be much more challenging.

While analyzing the interaction between control and communication within the MAC layer, the system was relatively simple; it was just an integrator system. Moreover, the impulsive input, which resets the state back to the origin when applied, is only realistic when the state represents an estimation error. This restriction was necessary for an exact mathematical analysis of the problem because, to the best of our knowledge, the interevent times of event-based control are only known for this kind of system. Considering more realistic systems would be very worthwhile and interesting. One step in this direction was done in Blind and Allgöwer (2012a), where the performance of event-based control of a scalar system with packet loss was studied. Here, one of the problems is the question how to increase the bounds after a packet loss. An alternative to an exact mathematical analysis of the interaction between control and communication on the MAC layer are simulations. Thus, event-based control of scalar systems with different MAC protocols have been compared in Högl (2010) by means of simulations.

Another interesting problem is the design of agents that autonomously choose the load. When an agent would be unaware of the fact that loss and delay depend on the load, it would conclude that the optimal performance is achieved with an infinite load. Obviously, such an agent would overload the communication system, which leads to a high loss probability and/or long delays and thus leads to a bad performance. A more intelligent agent could assume that all other agents keep their load constant and loss and delay is affected only by its own load. However, in Section 6.5 we showed that the cost of all agents together is minimal when all agents send equally. Thus, an intelligent agent must be aware of this fact and must be willing to share the available bandwidth fairly.

Within this thesis, we showed several methods to optimize the communication system to improve the control performance. In all presented approaches, only one layer of the ISO/OSI model was optimized. An even better control performance might be achieved by optimizing multiple layers of the ISO/OSI model. Since this might be very difficult, we should at least be aware of the most important effects, like the fact that loss and delay generally depend on the load. Within this thesis, the dependency between the design of the communication system, its load, loss, and delay and the consequences for the controller design were studied in detail in Chapter 6. We did not consider this dependency in detail in the other chapters. Nevertheless, we avoided to increase the load of the communication system because we were aware of the fact that loss and delay of a communication system generally increases with its load.

Taking into account that loss and delay depend on the load of the communication system would have the following consequences when the optimization of the transport layer is considered. When using TCP, the expected number of transmitted packets is smaller than the maximally allowed packets per sampling interval. In contrast, UDP always sends the maximally allowed number of packets. By using TCP, the load of the communication system will be smaller and thus a smaller loss and delay could be expected. Moreover, due to the acknowledgements in TCP, the controller implicitly senses packet loss and delay and could thus react to changes in the Quality of Service (QoS) of the communication system.

Moreover, taking into account that loss and delay depend on the load might lead to interesting scenarios when the route through the network is optimized periodically. Since a link with a low utilization has a low loss probability and a short delay, this link will be chosen on several routes. In doing so, the load of this link is increased, and, as a consequence thereof, also its loss and delay. In the next iteration, this link might be avoided, reducing its load and thus its loss and delay. Consequently, when optimizing the route periodically, we must be aware of such effects and work on methods to avoid resulting oscillations.

Appendix A.

Kalman Filtering with Intermittent Observations

In this section, we briefly repeat the results of Sinopoli et al. (2004). Although these results have been extended to the case of two and more channels in Liu and Goldsmith (2004b) and Garone et al. (2007), we stick to Sinopoli et al. (2004) for simplicity.

We consider the following discrete-time system

$$\mathbf{x}_{k+1} = A\mathbf{x}_k + \mathbf{w}_k, \tag{A.1a}$$

$$\mathbf{y}_k = \gamma_k C\mathbf{x}_k + \mathbf{v}_k, \tag{A.1b}$$

where $\mathbf{x}_k \in \mathbb{R}^{n_x}$ is the system state and $\mathbf{y}_k \in \mathbb{R}^{n_y}$ the measurement output at time instance k. Moreover, $\mathbf{w}_k \in \mathbb{R}^{n_x}$ and $\mathbf{v}_k \in \mathbb{R}^{n_y}$ are Gaussian white noise vectors with zero mean and covariance matrix $W \in \mathbb{R}^{n_x \times n_x}$ and $V \in \mathbb{R}^{n_y \times n_y}$, respectively. Moreover, $\gamma_k \in \{0, 1\}$ is an independent and identically distributed (iid) random process with $\mathrm{E}[\gamma_k] = p$, which indicates whether or not a measurement arrives.

First, we define

$$\hat{\mathbf{x}}_{k|k} := \mathrm{E}[\mathbf{x}_k | \mathcal{I}_k]$$
$$P_{k|k} := \mathrm{E}[(\mathbf{x}_k - \hat{\mathbf{x}}_k)(\mathbf{x}_k - \hat{\mathbf{x}}_k)^\mathsf{T} | \mathcal{I}_t]$$
$$\hat{\mathbf{x}}_{k+1|k} := \mathrm{E}[\mathbf{x}_{k+1} | \mathcal{I}_t]$$
$$P_{k+1|k} := \mathrm{E}[(\mathbf{x}_{k+1} - \hat{\mathbf{x}}_{k+1|k})(\mathbf{x}_{k+1} - \hat{\mathbf{x}}_{k+1|k})^\mathsf{T} | \mathcal{I}_k]$$

where $\mathcal{I}_k$ is the information available at time k, i.e. $\mathbf{y}_0, \dots, \mathbf{y}_k$ and $\gamma_0, \dots, \gamma_k$.

The time update of the Kalman filter is

$$\hat{\mathbf{x}}_{k+1|k} = A\hat{\mathbf{x}}_{k|k}, \tag{A.2a}$$

$$P_{k+1|k} = AP_{k|k}A^\mathsf{T} + W, \tag{A.2b}$$

which is identical to the case without measurement losses. The measurement update becomes

$$\hat{\mathbf{x}}_{k+1|k+1} = \hat{\mathbf{x}}_{k+1|k} + \gamma_{k+1}P_{k+1|k}C^\mathsf{T}\left(CP_{k+1|k}C^\mathsf{T} + V\right)^{-1}(\mathbf{y}_{t+1} - C\hat{\mathbf{x}}_{t+1|t}), \tag{A.3a}$$

$$P_{k+1|k+1} = P_{k+1|k} - \gamma_{k+1}P_{k+1|k}C^\mathsf{T}\left(CP_{k+1|k}C^\mathsf{T} + V\right)^{-1}CP_{k+1|}. \tag{A.3b}$$

111

Both $\hat{x}_{k+1|k+1}$ and $P_{k+1|k+1}$ are now random variables, depending on γ_{k+1}. Not surprisingly, a measurement update is only performed when a measurement arrives.

Using the shortcut $P_k := P_{k|k-1}$, (A.2b) and (A.3b) can be written as

$$P_{k+1} = AP_k A^\mathsf{T} + W - \gamma_k AP_k C^\mathsf{T} \left(CP_k C^\mathsf{T} + V\right)^{-1} CP_k A^\mathsf{T}. \qquad (A.4)$$

In order to derive an upper bound of $\mathrm{E}[P_k]$, the Modified Algebraic Riccati Equation (MARE) $g_p(X)$ is defined as follows:

$$g_p(X) = AXA^\mathsf{T} + W - pAXC^\mathsf{T} \left(CXC^\mathsf{T} + V\right)^{-1} CXA^\mathsf{T} \qquad (A.5)$$

After these definitions, Sinopoli et al. (2004) showed that there exists a critical arrival rate p_c which determines whether $\mathrm{E}[P_k]$ is bounded or not and gave an upper bound for p_c and $\mathrm{E}[P_k]$.

Theorem A.1 (Sinopoli et al. (2004)). *If $(A, W^{\frac{1}{2}})$ is controllable, (A, C) is detectable, and A is unstable, then there exists a $p_c \in [0, 1)$ such that*

$$\lim_{k \to \infty} \mathrm{E}[P_k] = +\infty, \qquad \text{for } 0 \leq p \leq p_c \text{ and } \exists P_0 \geq 0$$

$$\mathrm{E}[P_k] \leq M_{P_0} \, \forall k, \qquad \text{for } p_c < p \leq 1 \text{ and } \forall P_0 \geq 0$$

where $M_{P_0} > 0$ depends on the initial condition $P_0 \geq 0$.

Unfortunately, p_c can not be calculated directly, but an upper bound $\bar{p}_c$ can be found as follows:

Theorem A.2 (Sinopoli et al. (2004)). *The upper bound $\bar{p}_c$ is given by the solution of the following optimization problem*

$$\bar{p}_c = \arg\min_p \Psi(Y, Z) > 0, \qquad 0 \leq Y \leq I,$$

where

$$\Psi(Y, Z) = \begin{bmatrix} Y & \sqrt{p}(YA + ZC) & \sqrt{1-p}\,YA \\ \star & Y & 0 \\ \star & \star & Y \end{bmatrix}.$$

The following theorem shows that there also exists a lower and upper bound for $\mathrm{E}[P_k]$:

Theorem A.3 (Sinopoli et al. (2004)). *Assume that $(A, W^{\frac{1}{2}})$ is controllable, (A, C) is detectable and $p > \bar{p}_c$. Then*

$$0 < \underline{P}_k \leq \mathrm{E}[P_k] \leq \bar{P}_k \qquad \forall \mathrm{E}[P_0] \geq 0,$$

where $\underline{P}_k$ is found by the sequence $\underline{P}_{k+1} = (1-p)A\underline{P}_k A^\mathsf{T} + W$; $\underline{P}_0 = 0$ and $\bar{P}_k$ is found by the sequence $\bar{P}_{k+1} = g_p(\bar{P}_k)$, $\bar{P}_0 = E[P_0] \geq 0$. Moreover, $\lim_{k \to \infty} \underline{P}_k = \underline{P}_\infty$, where $\underline{P}_\infty$ is the solution of the algebraic equation $\underline{P}_\infty = (1 - p)A\underline{P}_\infty A^\mathsf{T} + W$. Similarly, $\lim_{k \to \infty} \bar{P}_k = \bar{P}_\infty$ where $\bar{P}_\infty$ is the fixed point of (A.5), i.e., $\bar{P}_\infty = g_p(\bar{P}_\infty)$.

The last theorem states that the upper bound $\bar{P}_\infty$ can be found by an LMI-problem.

Theorem A.4 (Sinopoli et al. (2004)). *If $p > \bar{p}_c$, then the matrix $\bar{P}_\infty = g_p(\bar{P}_\infty)$ is given by*

a) $\bar{P}_\infty = \lim_{k \to \infty} \bar{P}_k$; $\bar{P}_{k+1} = g_p(\bar{P}_k)$ *where* $\bar{P}_0 \geq 0$.

b) $\bar{P}_\infty = \arg\max_S \operatorname{Tr}\{S\}$

$$\text{subject to} \quad \begin{bmatrix} ASA^\mathsf{T} - S + W & \sqrt{p}ASC^\mathsf{T} \\ \star & CSC^\mathsf{T} + V \end{bmatrix} \geq 0, \ S \geq 0.$$

Appendix B.

Scale and Location Parameters of Probability Distributions

In this section, we discuss the location and scale parameter of a family of distributions, based on, e.g., Definition 1, 2, and 4 of Ferguson (1962) and Section 6.5.1 of Mukhopadhyay (2000).

Given a random variable z with PDF $f_z(z)$ and CDF $F_z(z)$, we can define a new random variable $x := az + b$ with PDF $f_x(x|a, b) = \frac{1}{a} f_z\left(\frac{x-b}{a}\right)$ and CDF $F_x(x|a, b) = F_z\left(\frac{x-b}{a}\right)$. In doing so, we get a family of distributions. Since the parameter a scales the random variable and b changes its location, these parameters are consequently called scale and location parameter. More formally, these parameters and families are defined as follows.

Definition B.1 (Ferguson (1962), Definition 1). *A real parameter θ is said to be a location parameter of a location family of distributions if $F(x|\theta)$ is a function only of $x - \theta$.*

If a density $f(x|\theta)$ exists, then θ is a location parameter if and only if $f(x|\theta) = g(x - \theta)$ for some function g, see Ferguson (1962).

Definition B.2 (Ferguson (1962), Definition 2). *A positive real parameter θ is said to be a scale parameter of a scale family of distributions if $F(x|\theta)$ is a function only of $x\theta^{-1}$.*

If a density $f(x|\theta)$ exists, then θ is a scale parameter if and only if $f(x|\theta) = \theta^{-1} g(x\theta^{-1})$ for some function g, see Ferguson (1962).

Definition B.3 (Ferguson (1962), Definition 4). *The tuple (μ, σ) with $\sigma > 0$ is said to be a location-scale parameter of a location-scale family of distributions $F(x|\mu, \sigma)$ if $F(x|\mu, \sigma)$ is a function only of $(x - \mu)\sigma^{-1}$.*

If a density $f(x|\mu, \sigma)$ exists, then (μ, σ) is a location-scale parameter if and only if $f(x|\mu, \sigma) = \sigma^{-1} g\left((x - \mu)\sigma^{-1}\right)$ for some function g, see Ferguson (1962).

Note that for $\mu \neq 0$, the parameter σ is not a scale parameter as defined in Definition B.2.

Definition B.4. *The term* rate parameter *is used to denote the inverse of the scale parameter.*

Example B.5 (Normal distribution). The PDF of the normal distribution with mean μ and standard deviation σ is

$$f(x|\mu,\sigma) = \frac{1}{\sigma\sqrt{2\pi}}e^{-\frac{(x-\mu)^2}{2\sigma^2}}.$$

We immediately see that this is a location-scale family. Moreover, with $\mu = 0$ and $\sigma = 1$, the standard normal distribution $f(x) = \frac{1}{\sqrt{2\pi}}e^{-\frac{1}{2}x^2}$ is obtained.

Example B.6 (Negative exponential distribution). The PDF of the negative exponential distribution is most often given with its rate parameter λ:

$$f(x|\lambda) = \begin{cases} \lambda e^{-\lambda x} & x \geq 0, \\ 0 & x < 0. \end{cases}$$

Alternatively, the negative exponential distribution can also be defined with the scale parameter β:

$$f(x|\beta) = \begin{cases} \frac{1}{\beta}e^{-x/\beta} & x \geq 0, \\ 0 & x < 0. \end{cases}$$

Appendix C.

Remaining Proofs of Chapter 5

C.1. Proof of Theorem 5.1

Theorem 5.1 is an extension of Theorem 4.6 from Kögel (2009) to allow a cross term in the cost function. Thus, the proof is similar to the one given in Kögel (2009), which is based on Koning (1992).

We define

$$\mathbf{z}_k := \begin{bmatrix} \mathbf{x}_k \\ \hat{\mathbf{x}}_k \end{bmatrix},$$

to write the closed loop system (5.5), (5.10) as

$$\mathbf{z}_{k+1} = \mathcal{A}_k \mathbf{z}_k + \tilde{\mathbf{w}}_k,$$

where

$$\mathcal{A}_k = \begin{bmatrix} A & -\beta_k BK \\ \gamma_k LC & A - E[\beta_k|\text{ACK}_k]BK - \gamma_k LC \end{bmatrix}, \qquad \tilde{\mathbf{w}}_k = \begin{bmatrix} \mathbf{w}_k \\ \gamma_k L\mathbf{v}_k \end{bmatrix}.$$

The covariance matrix of $\tilde{\mathbf{w}}_k$ is

$$\mathcal{W} = E[\tilde{\mathbf{w}}_k \tilde{\mathbf{w}}_k^\mathsf{T}] = \begin{bmatrix} W & 0 \\ 0 & \bar{p}_{sc}LVL^\mathsf{T} \end{bmatrix}.$$

The covariance matrix of $\mathbf{z}_k$ is

$$\mathcal{P}_k = E[\mathbf{z}_k \mathbf{z}_k^\mathsf{T}]$$

and evolves as

$$\mathcal{P}_{k+1} = E[\mathcal{A}_k \mathcal{P}_k \mathcal{A}_k^\mathsf{T}] + \mathcal{W}.$$

Moreover, the cost per step c_k is

$$c_k = \text{Tr}(\mathcal{N}_k \mathcal{P}_k), \qquad \mathcal{N}_k := \begin{bmatrix} Q & -\beta_k HK \\ -\beta_k K^\mathsf{T} H^\mathsf{T} & \beta_k K^\mathsf{T} RK \end{bmatrix}.$$

When $\mathcal{P}_k$ converges, we have

$$\mathcal{P}_\infty = E[\mathcal{A}\mathcal{P}_\infty \mathcal{A}^\mathsf{T}] + \mathcal{W}$$

and the expected cost per step is

$$c_\infty = \mathrm{Tr}(\mathcal{N}\mathcal{P}_\infty), \qquad \mathcal{N} = \mathrm{E}[\mathcal{N}_k].$$

Minimizing the expected cost per step gives the following optimization problem

$$\min \mathrm{Tr}(\mathcal{N}\mathcal{P}_\infty)$$
$$\text{s.t. } \mathrm{E}[\mathcal{A}\mathcal{P}_\infty\mathcal{A}^\mathsf{T}] + \mathcal{W} - \mathcal{P}_\infty = 0.$$

To obtain necessary conditions, we use the Lagrange multiplier Λ to get

$$\mathcal{L} = \mathrm{Tr}(\mathcal{N}\mathcal{P}_\infty + \Lambda\,\mathrm{E}[\mathcal{A}\mathcal{P}_\infty\mathcal{A}^\mathsf{T}] + \Lambda\mathcal{W} - \Lambda\mathcal{P}_\infty).$$

Using the matrix minimum principle (Athans (1967)) we get the following necessary conditions

$$\frac{\partial}{\partial\mathcal{P}_\infty}\mathcal{L} = \mathcal{N} + \mathrm{E}[\mathcal{A}^\mathsf{T}\Lambda\mathcal{A}] - \Lambda = 0 \tag{C.1}$$

$$\frac{\partial}{\partial\Lambda}\mathcal{L} = \mathcal{W} + \mathrm{E}[\mathcal{A}\mathcal{P}_\infty\mathcal{A}^\mathsf{T}] - \mathcal{P}_\infty = 0 \tag{C.2}$$

$$\frac{\partial}{\partial K}\mathcal{L} = \frac{\partial}{\partial K}\,\mathrm{Tr}\big((\mathcal{N} + \mathrm{E}[\mathcal{A}^\mathsf{T}\Lambda\mathcal{A}] - \Lambda)\mathcal{P}_\infty\big) = 0 \tag{C.3}$$

$$\frac{\partial}{\partial L}\mathcal{L} = \frac{\partial}{\partial L}\,\mathrm{Tr}\big(\Lambda(\mathcal{W} + \mathrm{E}[\mathcal{A}\mathcal{P}_\infty\mathcal{A}^\mathsf{T}] - \mathcal{P}_\infty)\big) = 0. \tag{C.4}$$

To simplify notation, we partition $\mathcal{P}_\infty$ and Λ as follows

$$\mathcal{P}_\infty = \begin{bmatrix} \underline{P}+\overline{P} & \underline{P} \\ \underline{P} & \underline{P} \end{bmatrix}, \qquad \Lambda = \begin{bmatrix} \underline{\Lambda}+\overline{\Lambda} & -\underline{\Lambda} \\ -\underline{\Lambda} & \underline{\Lambda} \end{bmatrix}$$

with $\underline{P} = \underline{P}^\mathsf{T} > 0$, $\overline{P} = \overline{P}^\mathsf{T} \geq 0$, $\underline{\Lambda} = \underline{\Lambda}^\mathsf{T} > 0$, and $\overline{\Lambda} = \overline{\Lambda}^\mathsf{T} \geq 0$ and define

$$\Xi := \mathcal{N} + \mathrm{E}[\mathcal{A}^\mathsf{T}\Lambda\mathcal{A}] - \Lambda, \qquad \Xi = \begin{bmatrix} \Xi_{11} & \Xi_{12} \\ \Xi_{21} & \Xi_{22} \end{bmatrix},$$

$$\Pi := \mathcal{W} + \mathrm{E}[\mathcal{A}\mathcal{P}\mathcal{A}^\mathsf{T}] - \mathcal{P}, \qquad \Pi = \begin{bmatrix} \Pi_{11} & \Pi_{12} \\ \Pi_{21} & \Pi_{22} \end{bmatrix}.$$

Using this notation, (C.1) - (C.4) can be written as

$$\text{(C.1)}: \quad \Xi = 0 \tag{C.5}$$

$$\text{(C.2)}: \quad \Pi = 0 \tag{C.6}$$

$$\text{(C.3)}: \quad \frac{\partial}{\partial K}\,\mathrm{Tr}\big((\Xi_{11} + \Xi_{12} + \Xi_{21} + \Xi_{22})\underline{P} + \Xi_{11}\overline{P}\big) = 0, \tag{C.7}$$

$$\text{(C.4)}: \quad \frac{\partial}{\partial L}\,\mathrm{Tr}\big(\underline{\Lambda}(\Pi_{11} - \Pi_{12} - \Pi_{21} + \Pi_{22}) + \overline{\Lambda}\Pi_{11}\big) = 0. \tag{C.8}$$

Straight forward calculations lead to

$$\Xi_{11} = \mathrm{E}\big[A^\mathsf{T}\overline{\Lambda}A + A^\mathsf{T}\underline{\Lambda}A - \gamma_k C^\mathsf{T}L^\mathsf{T}\underline{\Lambda}A - \gamma_k A^\mathsf{T}\underline{\Lambda}LC + \gamma_k^2 C^\mathsf{T}L^\mathsf{T}\underline{\Lambda}LC + Q - \overline{\Lambda} - \underline{\Lambda}\big],$$

$$\begin{aligned}\Xi_{12} = \mathrm{E}\big[&-\beta_k A^\mathsf{T}\overline{\Lambda}BK - \beta_k A^\mathsf{T}\underline{\Lambda}BK + \beta_k\gamma_k C^\mathsf{T}L^\mathsf{T}\underline{\Lambda}BK - A^\mathsf{T}\underline{\Lambda}A + \gamma_k C^\mathsf{T}L^\mathsf{T}\underline{\Lambda}A \\ &+ \gamma_k A^\mathsf{T}\underline{\Lambda}LC - \gamma_k^2 C^\mathsf{T}L^\mathsf{T}\underline{\Lambda}LC + \mathrm{E}[\beta_k|\mathrm{ACK}_k]A^\mathsf{T}\underline{\Lambda}BK \\ &- \gamma_k\,\mathrm{E}[\beta_k|\mathrm{ACK}_k]C^\mathsf{T}L^\mathsf{T}\underline{\Lambda}BK - \beta_k HK + \underline{\Lambda}\big],\end{aligned}$$

$$\begin{aligned}\Xi_{21} = \mathrm{E}\big[&-\beta_k K^\mathsf{T}B^\mathsf{T}\overline{\Lambda}A - \beta_k K^\mathsf{T}B^\mathsf{T}\underline{\Lambda}A - A^\mathsf{T}\underline{\Lambda}A + \gamma_k C^\mathsf{T}L^\mathsf{T}\underline{\Lambda}A + \mathrm{E}[\beta_k|\mathrm{ACK}_k]K^\mathsf{T}B^\mathsf{T}\underline{\Lambda}A \\ &+ \beta_k\gamma_k K^\mathsf{T}B^\mathsf{T}\underline{\Lambda}LC + \gamma_k A^\mathsf{T}\underline{\Lambda}LC - \gamma_k^2 C^\mathsf{T}L^\mathsf{T}\underline{\Lambda}LC - \gamma_k\,\mathrm{E}[\beta_k|\mathrm{ACK}_k]K^\mathsf{T}B^\mathsf{T}\underline{\Lambda}LC \\ &- \beta_k K^\mathsf{T}H^\mathsf{T} + \underline{\Lambda}\big],\end{aligned}$$

$$\begin{aligned}\Xi_{22} = \mathrm{E}\big[&\beta_k^2 K^\mathsf{T}B^\mathsf{T}\overline{\Lambda}BK + \beta_k^2 K^\mathsf{T}B^\mathsf{T}\underline{\Lambda}BK + \beta_k A^\mathsf{T}\underline{\Lambda}BK - \beta_k\gamma_k C^\mathsf{T}L^\mathsf{T}\underline{\Lambda}BK \\ &- 2\beta_k\,\mathrm{E}[\beta_k|\mathrm{ACK}_k]K^\mathsf{T}B^\mathsf{T}\underline{\Lambda}BK + \beta_k K^\mathsf{T}B^\mathsf{T}\underline{\Lambda}A + A^\mathsf{T}\underline{\Lambda}A - \gamma_k C^\mathsf{T}L^\mathsf{T}\underline{\Lambda}A \\ &- \mathrm{E}[\beta_k|\mathrm{ACK}_k]K^\mathsf{T}B^\mathsf{T}\underline{\Lambda}A - \beta_k\gamma_k K^\mathsf{T}B^\mathsf{T}\underline{\Lambda}LC - \gamma_k A^\mathsf{T}\underline{\Lambda}LC + \gamma_k^2 C^\mathsf{T}L^\mathsf{T}\underline{\Lambda}LC \\ &+ \gamma_k\,\mathrm{E}[\beta_k|\mathrm{ACK}_k]K^\mathsf{T}B^\mathsf{T}\underline{\Lambda}LC - \mathrm{E}[\beta_k|\mathrm{ACK}_k]A^\mathsf{T}\underline{\Lambda}BK \\ &+ \gamma_k\,\mathrm{E}[\beta_k|\mathrm{ACK}_k]C^\mathsf{T}L^\mathsf{T}\underline{\Lambda}BK + \mathrm{E}[\beta_k|\mathrm{ACK}_k]^2 K^\mathsf{T}B^\mathsf{T}\underline{\Lambda}BK + \beta_k K^\mathsf{T}RK - \underline{\Lambda}\big],\end{aligned}$$

$$\Pi_{11} = \mathrm{E}\big[A\overline{P}A^\mathsf{T} + A\underline{P}A^\mathsf{T} - \beta_k BK\underline{P}A^\mathsf{T} - \beta_k A\underline{P}K^\mathsf{T}B^\mathsf{T} + \beta_k^2 BK\underline{P}K^\mathsf{T}B^\mathsf{T} + W - \overline{P} - \underline{P}\big],$$

$$\begin{aligned}\Pi_{12} = \mathrm{E}\big[&\gamma_k A\overline{P}C^\mathsf{T}L^\mathsf{T} + A\underline{P}A^\mathsf{T} - \mathrm{E}[\beta_k|\mathrm{ACK}_k]A\underline{P}K^\mathsf{T}B^\mathsf{T} - \beta_k BK\underline{P}A^\mathsf{T} \\ &+ \beta_k\,\mathrm{E}[\beta_k|\mathrm{ACK}_k]BK\underline{P}K^\mathsf{T}B^\mathsf{T} - \underline{P}\big],\end{aligned}$$

$$\begin{aligned}\Pi_{21} = \mathrm{E}\big[&\gamma_k LC\overline{P}A^\mathsf{T} + A^\mathsf{T}\underline{P}A^\mathsf{T} - \mathrm{E}[\beta_k|\mathrm{ACK}_k]BK\underline{P}A^\mathsf{T} - \beta_k A\underline{P}K^\mathsf{T}B^\mathsf{T} \\ &+ \beta_k\,\mathrm{E}[\beta_k|\mathrm{ACK}_k]BK\underline{P}K^\mathsf{T}B^\mathsf{T} - \underline{P}\big],\end{aligned}$$

$$\begin{aligned}\Pi_{22} = \mathrm{E}\big[&\gamma_k^2 LC\overline{P}C^\mathsf{T}L^\mathsf{T} + A\underline{P}A^\mathsf{T} - \mathrm{E}[\beta_k|\mathrm{ACK}_k]BK\underline{P}A^\mathsf{T} - \mathrm{E}[\beta_k|\mathrm{ACK}_k]A\underline{P}K^\mathsf{T}B^\mathsf{T} \\ &+ \mathrm{E}[\beta_k|\mathrm{ACK}_k]^2 BK\underline{P}K^\mathsf{T}B^\mathsf{T} + \gamma_k LVL^\mathsf{T} - \underline{P}\big].\end{aligned}$$

To proceed, we derive some expected values that appear later on.

$$\overline{p}_{\mathrm{ca}} = \mathrm{E}\big[\mathrm{E}[\beta_k|\mathrm{ACK}_k]\big]$$

$$\begin{aligned}\overline{\epsilon} = \mathrm{E}[\beta_k|\mathrm{ACK}_k = 0] \\ = 1\cdot\frac{\Pr\{\beta_k = 1, \mathrm{ACK}_k = 0\}}{\Pr\{\mathrm{ACK}_k = 0\}} + 0\cdot\frac{\Pr\{\beta_k = 0, \mathrm{ACK}_k = 0\}}{\Pr\{\mathrm{ACK} = 0\}} = \frac{\overline{p}_{\mathrm{ca}}(1 - \overline{p}_{\mathrm{ac}})}{1 - \overline{p}_{\mathrm{ca}}\overline{p}_{\mathrm{ac}}}\end{aligned}$$

$$\sigma = \mathrm{E}\big[\mathrm{E}[\beta_k|\mathrm{ACK}_k]^2\big] = 1\cdot\Pr\{\mathrm{ACK}_k = 1\} + \overline{\epsilon}^2\cdot\Pr\{\mathrm{ACK}_k = 0\} = \overline{p}_{\mathrm{ca}}\overline{p}_{\mathrm{ac}} + \frac{\overline{p}_{\mathrm{ca}}^2(1 - \overline{p}_{\mathrm{ac}})^2}{1 - \overline{p}_{\mathrm{ca}}\overline{p}_{\mathrm{ac}}}$$

$$\begin{aligned}\overline{p}_{\mathrm{ca}}\phi = \mathrm{E}\big[\big(\beta_k - \mathrm{E}[\beta_k|\mathrm{ACK}_k]\big)^2\big] \\ = 0\cdot\Pr\{\beta_k = 1, \mathrm{ACK}_k = 1\} + (1 - \overline{\epsilon})^2\cdot\Pr\{\beta_k = 1, \mathrm{ACK}_k = 0\} \\ + \overline{\epsilon}^2\cdot\Pr\{\beta_k = 0, \mathrm{ACK}_k = 0\} = \cdots = \overline{p}_{\mathrm{ca}}\frac{(1 - \overline{p}_{\mathrm{ca}})(1 - \overline{p}_{\mathrm{ac}})}{1 - \overline{p}_{\mathrm{ca}}\overline{p}_{\mathrm{ac}}}\end{aligned}$$

From (C.7), we get

$$\mathrm{E}\big[-\gamma_k \underline{\Delta} A \overline{P} C^\mathsf{T} + \gamma_k \underline{\Delta} L(C\overline{P}C^\mathsf{T} + V)\big] = 0$$

and thereby (5.11). Similarly, from (C.8), we get

$$\mathrm{E}\big[\beta_k B^\mathsf{T}\overline{\Lambda}A - \beta_k H^\mathsf{T} + \beta_k^2 B^\mathsf{T}\overline{\Lambda}BK + (\beta_k - \mathrm{E}[\beta_k|\mathrm{ACK}_k])^2 B^\mathsf{T}\underline{\Delta}BK + \beta_k RK\big] = 0$$

and thereby (5.12). Moreover, (5.14) follows directly from $\Pi_{22} = 0$; (5.13) follows from $\Pi_{11} - \Pi_{12} - \Pi_{21} + \Pi_{22} = 0$ and (5.11). Similarly, (5.15) follows from $\Xi_{11} + \Xi_{12} + \Xi_{21} + \Xi_{22} = 0$ and (5.12); (5.16) follows from $\Xi_{12} + \Xi_{21} + \Xi_{22} = 0$ and (5.12).

Appendix D.

Remaining Proofs of Chapter 6

In Åström and Bernhardsson (2002), the expected cost and interevent time for controlling system (6.1) with an impulsive input is studied for $\sigma_i = 1$ and an ideal communication. In Rabi and Johansson (2009), this result is extended to the case of a communication system with random packet loss. In order to prove Theorem 6.3, 6.4, and 6.5, we have to extend these results to arbitrary σ_i and packet delay.

For simplicity of notation, we drop the subindex which indicates the agent whenever possible.

Lemma D.1. *Suppose, the expected time between two transmission attempts is T and the packet loss probability is q. Then the expected time between two successful transmissions T_S is*

$$\mathrm{E}[T_S] = \frac{T}{1-q}.$$

Proof.

$$\mathrm{E}[T_S] = \sum_{k=1}^{\infty}(1-q)q^{k-1}kT = T\frac{(1-q)}{q}\sum_{k=0}^{\infty}kq^k = \frac{T}{1-q}.$$

$\square$

Time-triggered and event-based control with packet loss has been studied in Rabi and Johansson (2009). In the considered setup packets are also delayed, so this effect must be taken into account. To show how delayed events affect the cost, we use $\tilde{x}$ to denote the state of the system with delayed events, whereas x remains the state of the system without delayed events. Figure D.1 shows an example of this case. Suppose an event is generated at time t_k. If the corresponding impulse is applied immediately, then the state will be reset to the origin. Due to the delay d_k, the impulse is applied at time $t_k + d_k$. Since the process continues in the meantime, the impulse will not reset the state to the origin. Thus, the cost of the system with delayed events will be different from the cost of the system without delayed events as stated in the following lemma.

Lemma D.2. *Suppose input (6.2) is used to control system (6.1). Then, the cost of the closed loop system with delayed events is $J + \sigma^2 d$, where J is the cost of the system without delay and d the expected delay.*

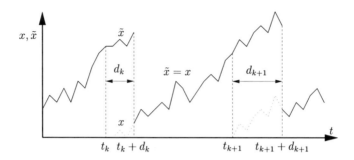

Figure D.1.: The effect of delayed events.

Proof. Since the interevent times are iid, it suffices to integrate between the two successful packet transmissions t_k and t_{k+1}. Thus, we get for the cost of the process with delayed events:

$$\tilde{J} = \frac{\mathrm{E}[\int_{t_k}^{t_{k+1}} \tilde{x}(t)^2 \mathrm{d}t]}{\mathrm{E}[t_{k+1} - t_k]}. \tag{D.1}$$

Obviously, we get for the denominator $\mathrm{E}[t_{k+1} - t_k] = \mathrm{E}[T_\mathrm{S}]$. Thus, we continue as follows:

$$
\begin{aligned}
\tilde{J} &= \frac{1}{\mathrm{E}[T_\mathrm{S}]} \mathrm{E}[\int_{t_k}^{t_{k+1}} \tilde{x}(t)^2 \mathrm{d}t] \\
&= \frac{1}{\mathrm{E}[T_\mathrm{S}]} \mathrm{E}[\int_{t_k}^{t_k+d_k} \tilde{x}(t)^2 \mathrm{d}t] + \frac{1}{\mathrm{E}[T_\mathrm{S}]} \mathrm{E}[\int_{t_k+d_k}^{t_{k+1}} \tilde{x}(t)^2 \mathrm{d}t] \\
&= \frac{1}{\mathrm{E}[T_\mathrm{S}]} \mathrm{E}[\int_{t_k}^{t_k+d_k} \big(x(t) + x(t_k)\big)^2 \mathrm{d}t] + \frac{1}{\mathrm{E}[T_\mathrm{S}]} \mathrm{E}[\int_{t_k+d_k}^{t_{k+1}} x(t)^2 \mathrm{d}t] \\
&= \underbrace{\frac{1}{\mathrm{E}[T_\mathrm{S}]} \mathrm{E}[\int_{t_k}^{t_{k+1}} x(t)^2 \mathrm{d}t]}_{J} + \underbrace{\frac{1}{\mathrm{E}[T_\mathrm{S}]} \mathrm{E}[\int_{t_k}^{t_k+d_k} x(t_k)^2 \mathrm{d}t]}_{J_{\mathrm{delay}}} + \underbrace{\frac{2}{\mathrm{E}[T_\mathrm{S}]} \mathrm{E}[\int_{t_k}^{t_k+d_k} x(t)x(t_k) \mathrm{d}t]}_{=J^*}.
\end{aligned}
$$

Note that $J^* = 0$ because $x(t)$ and $x(t_k)$ are independent and $\mathrm{E}[x(t)] = 0$. Thus, it remains to check $J_{\mathrm{delay}} = \frac{1}{\mathrm{E}[T_\mathrm{S}]} \int_{t_k}^{t_k+d_k} \mathrm{E}[x(t_k)^2] \mathrm{d}t$. Since the delay d_k is independent from $x(t_k)$ and $\mathrm{E}[d_k] = d$, we get for the additional cost due to delay $J_{\mathrm{delay}} = \frac{1}{\mathrm{E}[T_\mathrm{S}]} \mathrm{E}[x(t_k)^2]d$. Since the process was reset to the origin at the previous successful transmission t_{k-1}, we have $\mathrm{E}[x(t_k)^2] = \sigma^2 \mathrm{E}[T_\mathrm{S}]$ and consequently $J_{\mathrm{delay}} = \sigma^2 d$. □

D.1. Proof of Theorem 6.3

To prove Theorem 6.3, we first extend the cost of time-triggered control as given in Åström and Bernhardsson (2002) to an arbitrary noise intensity σ. Based on this result, we then derive the cost of time-triggered control with packet loss.

Lemma D.3. *Suppose, system* (6.1) *is controlled by an impulsive time-triggered control scheme with sampling time* T_{TT} *and an ideal communication without loss and delay. Then the cost is*

$$J_{\mathrm{TT}} = \sigma^2 \frac{T_{\mathrm{TT}}}{2}. \tag{D.2}$$

Proof.

$$J_{\mathrm{TT}} = \frac{1}{T_{\mathrm{TT}}} \mathrm{E}[\int_0^{T_{\mathrm{TT}}} x^2(t)\mathrm{d}t] = \frac{1}{T_{\mathrm{TT}}} \int_0^{T_{\mathrm{TT}}} \mathrm{E}[x^2(t)]\mathrm{d}t = \frac{1}{T_{\mathrm{TT}}} \int_0^{T_{\mathrm{TT}}} \sigma^2 t \mathrm{d}t = \sigma^2 \frac{T_{\mathrm{TT}}}{2}.$$

$\square$

Lemma D.4. *Suppose, system* (6.1) *is controlled by an impulsive time-triggered control scheme with sampling time* T_{TT} *and a packet loss probability* q. *Then the cost is*

$$J_{\mathrm{TT}} = \sigma^2 \left(\frac{T_{\mathrm{TT}}}{2} + \frac{T_{\mathrm{TT}}q}{(1-q)} \right).$$

Proof. To prove Lemma D.4, we use J_m to denote the cost for the case that the sampling time is extended to mT_{TT}, $m \in \mathbb{N}^+$. From Lemma D.3, we have

$$J_m = \frac{\mathrm{E}[\int_0^{mT_{\mathrm{TT}}} x^2(t)\mathrm{d}t]}{mT_{\mathrm{TT}}} = \sigma^2 \frac{mT_{\mathrm{TT}}}{2}$$

and thus,

$$\mathrm{E}[\int_0^{mT_{\mathrm{TT}}} x^2(t)\mathrm{d}t] = J_m mT_{\mathrm{TT}} = \sigma^2 \frac{m^2 T_{\mathrm{TT}}^2}{2}.$$

Now, we consider the term $\mathrm{E}[\int_0^{T_{\mathrm{S}}} x^2(t)\mathrm{d}t]$, where T_{S} is the expected time between two successful packet transmissions:

$$\mathrm{E}[\int_0^{T_{\mathrm{S}}} x^2(t)\mathrm{d}t] = \sum_{m=1}^{\infty} (1-q)q^{m-1} \mathrm{E}[\int_0^{mT_{\mathrm{TT}}} x^2(t)\mathrm{d}t]$$

$$= \frac{\sigma^2 T_{\mathrm{TT}}^2}{2} \frac{1-q}{q} \sum_{m=0}^{\infty} m^2 q^m = \frac{\sigma^2 T_{\mathrm{TT}}^2}{2} \frac{1+q}{(1-q)^2}.$$

Thus, we get for the cost

$$J_{\mathrm{TT}} = \frac{\mathrm{E}[\int_0^{T_{\mathrm{S}}} x^2(t)\mathrm{d}t]}{\mathrm{E}[T_{\mathrm{S}}]} = \frac{\sigma^2 T_{\mathrm{TT}}}{2} \frac{1+q}{1-q} = \sigma^2 \left(\frac{T_{\mathrm{TT}}}{2} + \frac{T_{\mathrm{TT}}q}{(1-q)} \right).$$

$\square$

Proof of Theorem 6.3. Theorem 6.3 follows from combining Lemma D.2 and D.4. $\square$

D.2. Proof of Theorem 6.4

To prove Theorem 6.4, we use a result from Feller (1954).

Lemma D.5 (Feller (1954)). *Suppose, the lower bound is $\underline{\Delta}$ and the upper bound is $\overline{\Delta}$ and system (6.1) is started at x_0; $\underline{\Delta} < x_0 < \overline{\Delta}$. Then, the expected time to reach one of these bounds is characterized as the unique solution of*

$$\frac{1}{2}\sigma^2 \frac{\partial^2}{\partial x^2}\Psi(x_0) = -1 \qquad \text{with } \Psi(\underline{\Delta}) = \Psi(\overline{\Delta}) = 0. \tag{D.3}$$

Proof of Theorem 6.4. For the considered event-based control, we have $\underline{\Delta} = -\Delta, \overline{\Delta} = +\Delta$. Thus, the solution of (D.3) is

$$\Psi(x_0) = \frac{1}{\sigma^2}(\Delta^2 - x_0).$$

Now, note that in the considered setup, the system is reset to the origin, i.e., $x_0 = 0$. Thus, the expected interevent time is

$$T_{\text{EB}} = \Psi(0) = \frac{\Delta^2}{\sigma^2}.$$

$\square$

D.3. Proof of Theorem 6.5

Again, we extend the cost given in Åström and Bernhardsson (2002) to the case of an arbitrary noise intensity σ. Based on this result, we derive the cost of event-based control with packet loss similar to Rabi and Johansson (2009).

Lemma D.6. *Suppose, system (6.1) is controlled by an impulsive event-based control scheme with boundary increment Δ and an ideal communication without loss and delay. Then the cost is*

$$J_{\text{EB}} = \sigma^2 \frac{T_{\text{EB}}}{6}. \tag{D.4}$$

Proof. The distribution of the state follows from the steady state of the Kolmogorov forward equation

$$0 = \frac{1}{2}\frac{\partial^2}{\partial x^2}\sigma^2 f(x) \qquad \text{with } \int_{-\Delta}^{\Delta} f(x)\mathrm{d}x = 1, \, f(-\Delta) = f(\Delta) = 0.$$

This equation has the solution

$$f(x) = \frac{\Delta - |x|}{\Delta^2}.$$

Consequently, the variance is

$$J_{\text{EB}} = \int_{-\Delta}^{\Delta} x^2 f(x) \mathrm{d}x = \frac{\Delta^2}{6} = \sigma^2 \frac{T_{\text{EB}}}{6}.$$

$\square$

Lemma D.7. *Suppose, system (6.1) is controlled by an impulsive event-based control scheme with boundary increment Δ and a packet loss probability q. Then the cost is*

$$J_{\text{EB}} = \sigma^2 \left(\frac{T_{\text{EB}}}{6} + \frac{T_{\text{EB}}q}{(1-q)} \right). \tag{D.5}$$

Proof. To prove Lemma D.7, we use T_m to denote the time between m transmission attempts and T_s the time between two successful transmissions. We start by considering the term $\mathrm{E}[\int_0^{T_s} x^2(t)\mathrm{d}t]$

$$\mathrm{E}\left[\int_0^{T_s} x^2(t)\mathrm{d}t \right] = \sum_{m=1}^{\infty} (1-q)q^{m-1} \mathrm{E}\left[\int_0^{T_m} x^2(t)\mathrm{d}t \right] = \frac{1-q}{q} \sum_{m=0}^{\infty} q^m \sum_{n=1}^{m} \mathrm{E}\left[\int_{T_{n-1}}^{T_n} x^2(t)\mathrm{d}t \right].$$

To proceed, we consider the term $\nu_n := \mathrm{E}[\int_{T_{n-1}}^{T_n} x^2(t)\mathrm{d}t]$:

$$\nu_n := \mathrm{E}\left[\int_{T_{n-1}}^{T_n} x^2(t)\mathrm{d}t \right] = \mathrm{E}[x^2(T_{n-1})] \int_{T_{n-1}}^{T_n} \mathrm{d}t + \int_{T_{n-1}}^{T_n} (x(t) - x(T_{n-1}))^2 \mathrm{d}t].$$

Now, note that $x(T_n)$ is always an integer multiple of Δ and similar to the random process $\sum_{m=1}^{n} \theta(m)\Delta$, where $\theta(m) \in \{-1, 1\}$ with $\Pr\{\theta(m) = -1\} = 1/2$ and $\Pr\{\theta(m) = 1\} = 1/2$. Thus, ν_n becomes

$$\nu_n = \mathrm{E}\left[\left(\sum_{m=1}^{n} \theta_m \Delta \right)^2 \right] \mathrm{E}[T_{\text{EB}}] + \mathrm{E}\left[\int_0^{T_{\text{EB}}} x^2(t)\mathrm{d}t | x(0) = 0 \right] = (n-1)\frac{\Delta^4}{\sigma^2} + \frac{\Delta^4}{6\sigma^2}.$$

Now, we proceed with the term $\mathrm{E}[\int_0^{T_s} x^2(t)\mathrm{d}t]$

$$\mathrm{E}\left[\int_0^{T_s} x^2(t)\mathrm{d}t \right] = \frac{1-q}{q} \sum_{m=0}^{\infty} q^m \sum_{n=1}^{m} \nu_n$$

$$= \frac{1-q}{q} \sum_{m=0}^{\infty} q^m \sum_{n=1}^{m} (n-1)\frac{\Delta^4}{\sigma^2} + \frac{\Delta^4}{6\sigma^2}$$

$$= \frac{1-q}{q} \frac{\Delta^4}{\sigma^2} \sum_{m=0}^{\infty} q^m \left(\frac{m(m-1)}{2} + \frac{m}{6} \right)$$

$$= \frac{\Delta^4}{\sigma^2} \left(\frac{q}{(1-q)^2} + \frac{1}{6(1-q)} \right).$$

Finally, Lemma D.7 follows from $J_{\text{EB},i} = \frac{\mathrm{E}[\int_0^{T_s} x^2(t)\mathrm{d}t]}{\mathrm{E}[T_s]}$.

$\square$

Proof of Theorem 6.5. Theorem 6.5 follows from combining Lemma D.2 and D.7. $\square$

D.4. Proof of Theorem 6.9

To show that the arrival process of event-based control converges to a Poisson process for $N \to \infty$, we use the Palm-Khintchine Theorem and the following two assumptions, definition and short discussion, from Heyman and Sobel (1982).

Assumption D.8. *For all N sufficiently large,*

$$\lambda_{1,N} + \cdots + \lambda_{N,N} = \lambda_\Sigma < \infty, \tag{D.6}$$

where $\lambda_{j,N}$ is the sending rate of agent j for the case that there are N agents.

Assumption D.9. *Given $\epsilon > 0$, for each $t > 0$ and N sufficiently large,*

$$F_{j,N}(t) \leq \epsilon, \qquad j = 1, \ldots, N, \tag{D.7}$$

where $F_{j,N}$ is the Cumulative Distribution Function (CDF) of agent j for the case that there are N agents.

Definition D.10. *For each N define*

$$L_{0,N}(t) = L_{1,N}(t) + \cdots + L_{N,N}(t), \tag{D.8}$$

where $L_{j,N}(t)$ is a stochastic process, which counts the number of events of agent j that occur by time t for the case that there are N agents.

Assumption D.9 asserts that as N increases, the processes being combined have renewals very infrequently. Assumption D.8 shows that $L_{0,N+1}(t)$ is not formed by adding another process to $L_{0,N}(t)$. As N increases, the processes being combined are changed so that (at least for large N) the asymptotic rate at which renewals occur is a constant.

Theorem D.11 (Palm-Khintchine Theorem, (Heyman and Sobel, 1982, Theorem 5.15)). *Under Assumptions D.8 and D.9, as $N \to \infty$, $\{L_{0,N}(t); t \geq 0\}$ approaches a Poisson process.*

Proof of Theorem 6.9. Since Assumption D.8 is part of Theorem 6.9, it remains to show that Assumption D.9 holds. From the scaling property and the definition of the CDF, it follows that $F_{j,N}(t) = \int_0^t f(x|\Delta)dx = \int_0^t \lambda f(\lambda x|1)dx = \lambda_\Sigma/N \int_0^t f(\lambda_\Sigma x/N|1)dx$. Since $f(x|1)$ is continuous, if follows from the mean value theorem that $\int_0^t f(z/N|1)dx = f(z/N|1)t$ for some $z \in (0,t)$. Moreover, since $f(z/N) \to 0$ for $N \to \infty$, we see that for each $\epsilon > 0$ and $t > 0$ there exists an N such that $F_{j,N}(t) \leq \epsilon$, i.e., Assumption D.9 holds. $\square$

D.5. Proof of Theorem 6.10

For simplicity of notation we drop the subindex indicating the agent and use it to indicate the scheme instead. Moreover, we use $\tilde{q}$ to denote the probability of an unsuccessful transmission and $\tilde{d}$ for the delay between a packet transmission and its reception.

When using Scheme 1, all packets that are not successfully transmitted are lost and the delay is only the delay between event generation and the reception of the corresponding packet. Thus, for Scheme 1, the additional cost due to loss and delay is

$$\frac{\tilde{q}_1}{1 - \tilde{q}_1} T + \tilde{d}_1. \tag{D.9}$$

When using Scheme 2, all packets that are not transmitted successfully are retransmitted at some later time. Thus, there is no packet loss but an additional delay due to the retransmissions.

$$d_2 = \tilde{d}_2 + 0T(1 - \tilde{q}_2) + 1T(1 - \tilde{q}_2)\tilde{q}_2 + 2T(1 - \tilde{q}_2)\tilde{q}_2^2 + 3T(1 - \tilde{q}_2)\tilde{q}_2^3 + \cdots$$
$$= \tilde{d}_2 + (1 - \tilde{q}_2)\tilde{q}_2 T \sum_{m=0}^{\infty} m\tilde{q}_2^{m-1} = \tilde{d}_2 + \frac{\tilde{q}_2}{1 - \tilde{q}_2} T.$$

Thus, for Scheme 2, the additional cost due to delay becomes

$$\frac{\tilde{q}_2}{1 - \tilde{q}_2} T + \tilde{d}_2. \tag{D.10}$$

Now, we can compare the additional cost due to loss and delay of Scheme 1 and Scheme 2, given by (D.9) and (D.10). Due to the retransmission of packets, the load when using Scheme 2 is larger than the load when using Scheme 1. Since we assumed that either the probability of an unsuccessful transmission or delay, or both, is strictly increasing with the network load, Scheme 1 gives a better performance than Scheme 2. □

D.6. Proof of Lemma 6.14

We start by bounding the probability of self-interference, i.e., the term $(1 - \int_0^{\rho_i} f_i(x|1)dt)^2$. Since $f_i(t|1)$ is a PDF, $0 \leq 1 - \int_0^x f_i(t|1)dt \leq 1$ holds. From the assumption $f_i(t|1) \leq 1$, we also have $1 - x \leq 1 - \int_0^x f_i(t|1)dt$, i.e.

$$\max\{0, 1 - x\} \leq 1 - \int_0^x f_i(t|1)dt \leq 1. \tag{D.11}$$

Since $(1 - \rho_i)^2 = 1 - 2\rho_i + \rho_i^2 \leq 1 - 2\rho_i$, we can bound the probability of self-interference as follows

$$b_p(\rho_i) \leq \left(1 - \int_0^{\rho_i} f_i(x|1)dt\right)^2 \leq 1, \tag{D.12}$$

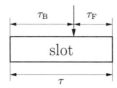

Figure D.2.: Forward and backward recurrence time.

with $b_p(\rho)$ as defined in (6.21).

Now, we continue with the probability of a collision with another user, i.e., the term $q_o(\rho_j) := 1 - \int_0^{2\rho_j}[1 - \int_0^x f_j(t|1)\mathrm{d}t]\mathrm{d}x$. Since $q_o(\rho_j)$ is a probability, we immediately have $0 \le q_o(\rho_j) \le 1$. Moreover, from (D.11) we get $1 - 2\rho_j \le q_o(\rho_j) \le 1 - 2\rho_j + 2\rho_j^2$. Since $q_o(\rho_j)$ is increasing with ρ_j and the upper bound $1 - 2\rho_j + 2\rho_j^2$ has a minimum at $\rho_j = 1/2$, we use $a_p(\rho_j)$ as upper bound for $q_o(\rho_j)$ for $\rho_j > 1/2$. Therefore, we get

$$b_p(\rho_j) \le 1 - \int_0^{2\rho_j}[1 - \int_0^x f_j(t|1)\mathrm{d}t]\mathrm{d}x \le a_p(\rho_j), \qquad (D.13)$$

with $a_p(\rho)$ and $b_p(\rho)$ as defined in (6.20) and (6.21), respectively.

Finally, (6.19) follows by using (D.12) and (D.13) in (6.18). □

D.7. Proof of Theorem 6.18

We start by proving the case without self-interference. Eq. (6.27) follows from (6.17) due to the fact that the vulnerable period is τ for slotted ALOHA, instead of 2τ for pure ALOHA.

Next, we analyze the losses due to self-interference. Figure D.2 shows an event generation within a slot and the corresponding times. The time interval between the event generation and the end of the slot is called *forward recurrence time* τ_F. Similarly, the time interval between the event generation and the begin of the slot is denoted *backward recurrence time* τ_B. As already stated, a self-interference occurs if the same user generates a new event while its packet is still waiting for the start of the next slot, i.e., during τ_F. Thus, the steady-state probability that a packet is not lost due to self-interference is

$$\mathrm{E}\left[\left(1 - \int_0^{\tau_F} f_i(t|\lambda_i)\mathrm{d}t\right)\right] = \frac{1}{\tau}\int_0^\tau\left(1 - \int_0^{\tau-x} f_i(t|\lambda_i)\mathrm{d}t\right)\mathrm{d}x. \qquad (D.14)$$

Finally, (6.26) follow from the fact that a packet of user i is lost if it interferes with a packet of user i or with a packet of any other user. □

D.8. Proof of Lemma 6.21

First, we look at the losses due to self-interference. Since the corresponding term is a probability and the assumption $f_i(x|1) \leq 1$, we get

$$b_s(\rho) \leq \frac{1}{\rho_i} \int_0^{\rho_i} \left(1 - \int_0^{\rho_i - x} f(t|1)\mathrm{d}t\right) \mathrm{d}x \leq 1. \qquad (D.15)$$

By following the same arguments as in the proof of Lemma 6.14, we get

$$b_s(\rho_j) \leq 1 - \int_0^{\rho_j} [1 - \int_0^x f_j(t|1)\mathrm{d}t]\mathrm{d}x \leq a_s(\rho_j). \qquad (D.16)$$

Finally, (6.29) follows by using (D.15) and (D.16) in (6.28). $\qquad \square$

Bibliography

N. Abramson. The ALOHA system – Another alternative for computer communications. In *Proceedings of the Fall Joint Computer Conference*, pages 281–286, 1970.

N. Abramson. The ALOHANet – Surfing for wireless data. *IEEE Communications Magazine*, 47(12):21–25, 2009.

A. Afanasyev, N. Tilley, P. Reiher, and L. Kleinrock. Host-to-host congestion control for TCP. *IEEE Communications Surveys and Tutorials*, 12(3):304–342, 2010.

D. Aldous. Ultimate instability of exponential back-off protocol for acknowledgment-based transmission control of random access communication channels. *IEEE Transactions on Information Theory*, 33(2):219–223, 1987.

A. Anta and P. Tabuada. To sample or not to sample: Self-triggered control for nonlinear systems. *IEEE Transactions on Automatic Control*, 55(9):2030–2042, 2010.

K. J. Åström and B. M. Bernhardsson. Comparison of Riemann and Lebesgue sampling for first order stochastic systems. In *Proceedings of the 41st IEEE Conference on Decision and Control (CDC)*, pages 2011–2016, Las Vegas, NV, USA, 2002.

M. Athans. The matrix minimum principle. *Information and Control*, 11(5-6):592–606, 1967.

R. Blind and F. Allgöwer. On the optimal sending rate for networked control systems with a shared communication medium. In *Proceedings of the 50th IEEE Conference on Decision and Control (CDC) and European Control Conference (ECC)*, pages 4704–4709, Orlando, FL, USA, 2011a.

R. Blind and F. Allgöwer. Analysis of networked event-based control with a shared communication medium: Part I – Pure ALOHA. In *Proceedings of the 18th IFAC World Congress*, pages 10092–10097, Milan, Italy, 2011b.

R. Blind and F. Allgöwer. Analysis of networked event-based control with a shared communication medium: Part II – Slotted ALOHA. In *Proceedings of the 18th IFAC World Congress*, pages 8830–8835, Milan, Italy, 2011c.

R. Blind and F. Allgöwer. The performance of event-based control for scalar systems with packet losses. In *Proceedings of the 51st IEEE Conference on Decision and Control (CDC)*, pages 6572–6576, Maui, HI, USA, 2012a.

R. Blind and F. Allgöwer. Is it worth to retransmit lost packets in networked control systems? In *Proceedings of the 51st IEEE Conference on Decision and Control (CDC)*, pages 1368–1373, Maui, HI, USA, 2012b.

R. Blind and F. Allgöwer. Retransmitting lost measurements to improve remote estimation. In *Proceedings of the American Control Conference (ACC)*, pages 4154–4158, Washington, DC, USA, 2013a.

R. Blind and F. Allgöwer. On time-triggered and event-based control of integrator systems over a shared communication system. *Mathematics of Control, Signals, and Systems*, 25(4):517–557, 2013b.

R. Blind and F. Allgöwer. On the joint design of controller and routing for networked control systems. In *Proceedings of the 4th IFAC Workshop on Distributed Estimation and Control in Networked Systems (NecSys)*, pages 240–246, Koblenz, Germany, 2013c.

R. Blind and F. Allgöwer. On the optimization of the transport layer for networked control systems. *at-Automatisierungstechnik*, 61(7):495–505, 2013d.

R. Blind and F. Allgöwer. On the stabilizability of continuous-time systems over a packet based communication system with loss and delay. *Accepted for 19th IFAC World Congress*, 2014.

R. Blind, S. Uhlich, B. Yang, and F. Allgöwer. Robustification and optimization of a Kalman filter with measurement loss using linear precoding. In *Proceedings of the American Control Conference (ACC)*, pages 2222–2227, St. Louis, MO, USA, 2009.

O. Brun and J.-M. Garcia. Analytical solution of finite capacity M/D/1 queues. *Journal of Applied Probablity*, 37(4):1092–1098, 2000.

B. W. Carabelli, A. Benzing, F. Dürr, B. Koldehofe, K. Rothermel, G. S. Seyboth, R. Blind, M. Bürger, and F. Allgöwer. Exact convex formulations of network-oriented optimal operator placement. In *Proceedings of the 51st IEEE Conference on Decision and Control (CDC)*, pages 3777–3782, Maui, HI, USA, 2012.

A. Cervin and T. Henningsson. Scheduling of event-triggered controllers on a shared network. In *Proceedings of the 47th IEEE Conference on Decision and Control (CDC)*, pages 3601–3606, Cancun, Mexico, 2008.

M. H. A. Davis and J. M. Howl. A Markovian analysis of the finite-buffer M/D/1 queue. *Proceedings of the Royal Society of London. Series A: Mathematical, Physical and Engineering Sciences*, 453(1964):1947–1962, 1997.

B. Demirel, Z. Zou, P. Soldati, and M. Johansson. Modular co-design of controllers and transmission schedules in WirelessHART. In *Proceedings of the 50th IEEE Conference on Decision and Control (CDC) and European Control Conference (ECC)*, pages 5951–5958, Orlando, FL, USA, 2011.

M. Epstein, L. Shi, A. Tiwari, and R. M. Murray. Probabilistic performance of state estimation across a lossy network. *Automatica*, 44(12):3046–3053, 2008.

A. W. Erlang. Solution of some problems in the theory of probabilities of significance in automatic telephone exchanges. *Post Office Electrical Engineer's Journal*, 10: 189–197, 1917.

W. Feller. *An Introduction to Probability Theory and its Applications*. John Wiley & Sons, New York, 1 edition, 1950.

W. Feller. Diffusion processes in one dimension. *Transactions of the American Mathematical Society*, 77(1):1–31, 1954.

T. S. Ferguson. Location and scale parameters in exponential families of distributions. *The Annals of Mathematical Statistics*, 33(3):986–1001, 1962.

G. F. Franklin, J. D. Powell, and M. Workman. *Digital Control of Dynamic Systems*. Addison-Wesley Longman Publishing Co., Inc., Boston, MA, USA, 3rd edition, 1997.

E. Garone, B. Sinopoli, A. Goldsmith, and A. Casavola. LQR control for distributed systems over TCP-like erasure channels. In *Proceedings of the 46th IEEE Conference on Decision and Control (CDC)*, pages 44–49, New Orleans, LA, USA, 2007.

E. Garone, B. Sinopoli, and A. Casavola. LQG control over lossy TCP-like networks with probabilistic packet acknowledgements. In *Proceedings of the 47th IEEE Conference on Decision and Control (CDC)*, pages 2686–2691, Cancun, Mexico, 2008.

V. K. Goyal. Multiple description coding: Compression meets the network. *IEEE Signal Processing Magazine*, 18(5):74–93, 2001.

V. K. Goyal and J. Kovacevic. Generalized multiple description coding with correlating transforms. *IEEE Transactions on Information Theory*, 47(6):2199–2224, 2001.

V. Gupta. On estimation across analog erasure links with and without acknowledgements. *IEEE Transactions on Automatic Control*, 55(12):2896–2901, 2010.

V. Gupta, A. F. Dana, J. P. Hespanha, R. M. Murray, and B. Hassibi. Data transmission over networks for estimation and control. *IEEE Transactions on Automatic Control*, 54(8):1807–1819, 2009.

W. P. M. H. Heemels, J. H. Sandee, and P. P. J. Van Den Bosch. Analysis of event-driven controllers for linear systems. *International Journal of Control*, 81(4):571–590, 2008.

T. Henningsson and A. Cervin. A simple model for the interference between event-based control loops using a shared medium. In *Proceedings of the 49th IEEE Conference on Decision and Control (CDC)*, pages 3240–3245, Atlanta, GA, USA, 2010.

T. Henningsson, E. Johannesson, and A. Cervin. Sporadic event-based control of first-order linear stochastic systems. *Automatica*, 44(11):2890–2895, 2008.

D. P. Heyman and M. J. Sobel. *Stochastic Models in Operations Research*, volume I. McGraw-Hill Book Company, New York, 1982.

J. Högl. *Simulation of networked event-based control*. Student thesis, Institute for Systems Theory and Automatic Control, University of Stuttgart, Germany, 2010.

D. Hristu-Varsakelis and P. Kumar. Interrupt-based feedback control over a shared communication medium. In *Proceedings of the 41st IEEE Conference on Decision and Control (CDC)*, pages 3223–3228, Las Vegas, NV, USA, 2002.

O. Imer, S. Yüksel, and T. Basar. Optimal control of LTI systems over unreliable communication links. *Automatica*, 42(9):1429–1439, 2006.

V. Jacobson. Congestion avoidance and control. *ACM SIGCOMM Computer Communication Review*, 18(4):314–329, 1988.

Z. Jin, V. Gupta, and R. M. Murray. State estimation over packet dropping networks using multiple description coding. *Automatica*, 42(9):1441–1452, 2006.

R. E. Kalman, Y. C. Ho, and K. S. Narendra. Controllability of linear dynamical systems. In *Contributions to Differential Equations*, volume 1, pages 189–213. Interscience, New York, 1963.

F. Kelly. Stochastic models of computer communication systems. *Journal of the Royal Statistical Society. Series B (Methodological)*, 47(3):379–395, 1985.

L. Kleinrock. *Queueing Systems: Volume I – Theory*. Wiley Insterscience, New York, 1975.

L. Kleinrock and S. S. Lam. Packet-switching in a slotted satellite channel. In *National Computer Conference, AFIPS Conference Proceedings*, pages 703–710, New York, NY, USA, 1973.

L. Kleinrock and F. Tobagi. Packet switching in radio channels: Part I – Carrier sense multiple-access modes and their throughput-delay characteristics. *IEEE Transactions on Communications*, 23(12):1400–1416, 1975.

E. Kofman and J. H. Braslavsky. Level crossing sampling in feedback stabilization under data-rate constraints. In *Proceedings of the 45th IEEE Conference on Decision and Control (CDC)*, pages 4423–4428, San Diego, CA, USA, 2006.

M. J. Kögel. On optimal control over networks with lossy links. Diploma thesis, Institute for Systems Theory and Automatic Control, University of Stuttgart, Germany, 2009. URL http://elib.uni-stuttgart.de/opus/volltexte/2010/5106.

M. J. Kögel, R. Blind, and F. Allgöwer. Optimal control over unreliable networks with uncertain loss rates. In *Proceedings of the American Control Conference (ACC)*, pages 3672–3677, Baltimore, MD, USA, 2010.

W. L. D. Koning. Compensatability and optimal compensation of systems with white parameters. *IEEE Transactions on Automatic Control*, 37(5):579–588, 1992.

S. S. Lam and L. Kleinrock. Packet switching in a multiaccess broadcast channel: Dynamic control procedures. *IEEE Transactions on Communications*, 23(9):891–904, 1975a.

S. S. Lam and L. Kleinrock. Dynamic control schemes for a packet switched multiaccess broadcast channel. In *Proceedings of the National Computer Conference*, pages 143–153, Anaheim, CA, USA, 1975b.

A. S. Leong, S. Dey, and J. S. Evans. On Kalman smoothing with random packet loss. *IEEE Transactions on Signal Processing*, 56(7):3346–3351, 2008.

A. H. Levis, R. A. Schlueter, and M. Athans. On the behaviour of optimal linear sampled-data regulators. *International Journal of Control*, 13(2):343–361, 1971.

X. Liu and A. Goldsmith. Wireless medium access control in networked control systems. In *Proceedings of the American Control Conference (ACC)*, pages 3605–3610, Boston, MA, USA, 2004a.

X. Liu and A. Goldsmith. Kalman filtering with partial observation losses. In *Proceedings of the 43rd IEEE Conference on Decision and Control (CDC)*, pages 4180–4186, Paradise Island, Bahamas, 2004b.

J. Lunze and D. Lehmann. A state-feedback approach to event-based control. *Automatica*, 46(1):211–215, 2010.

M. Mazo and P. Tabuada. Decentralized event-triggered control over wireless sensor/actuator networks. *IEEE Transactions on Automatic Control*, 56(10):2456–2461, 2011.

A. R. Mesquita, J. P. Hespanha, and G. N. Nair. Redundant data transmission in control/estimation over wireless networks. In *Proceedings of the American Control Conference (ACC)*, pages 3378–3383, St. Louis, MO, USA, 2009.

A. R. Mesquita, J. P. Hespanha, and G. N. Nair. Redundant data transmission in control/estimation over lossy networks. *Automatica*, 48(8):1612–1620, 2012.

A. Molin and S. Hirche. Optimal design of decentralized event-triggered controllers for large-scale systems with contention-based communication. In *Proceedings of the 50th IEEE Conference on Decision and Control (CDC) and European Control Conference (ECC)*, pages 4710–4716, Orlando, FL, USA, 2011.

N. Mukhopadhyay. *Probability and Statistical Inference*. CRC Press, 2000.

G. N. Nair, A. R. Mesquita, and J. P. Hespanha. Optimal redundant transmission for state estimation with packet drops. In *Proceedings of the 2nd IFAC Workshop on Distributed Estimation and Control in Networked Systems (NecSys)*, pages 163–168, Annecy, France, 2010.

D. J. Navarro and I. G. Fuss. Fast and accurate calculations for first-passage times in Wiener diffusion models. *Journal of Mathematical Psychology*, 53(4):222–230, 2009.

D. E. Quevedo, K. H. Johansson, A. Ahlén, and I. Jurado. Dynamic controller allocation for control over erasure channels. In *Proceedings of the 3rd IFAC Workshop on Distributed Estimation and Control in Networked Systems (NecSys)*, pages 61–66, Santa Barbara, CA, USA, 2012.

M. Rabi and K. H. Johansson. Scheduling packets for event-triggered control. In *Proceedings of the European Control Conference (ECC)*, pages 3779–3784, Budapest, Hungary, 2009.

M. Rabi and L. Stabellini. Analysis of networked estimation under contention-based medium access. In *Proceedings of the 17th IFAC World Congress*, pages 10283–10288, Seoul, Korea, 2008.

M. Rabi, G. V. Moustakides, and J. S. Baras. Adaptive sampling for linear state estimation. *SIAM Journal on Control and Optimization*, 50(2):672–702, 2012.

C. Ramesh, H. Sandberg, and K. H. Johansson. Steady state performance analysis of multiple state-based schedulers with CSMA. In *Proceedings of the 51st IEEE Conference on Decision and Control (CDC)*, pages 4729–4734, Maui, HI, USA, 2012a.

C. Ramesh, H. Sandberg, and K. H. Johansson. Stability analysis of multiple state-based schedulers with CSMA. In *Proceedings of the 51st IEEE Conference on Decision and Control (CDC)*, pages 7205–7211, Maui, HI, USA, 2012b.

C. Ramesh, H. Sandberg, and K. H. Johansson. Design of state-based schedulers for a network of control loops. *IEEE Transactions on Automatic Control*, 58(8):1962–1975, 2013.

L. Roberts. ALOHA packet system with and without slots and capture. *ACM SIGCOMM Computer Communication Review*, 5(2):28–42, 1975.

C. L. Robinson and P. R. Kumar. Optimizing controller location in networked control systems with packet drops. *IEEE Journal on Selected Areas in Communications*, 26(4):661–671, 2008.

R. Rom and M. Sidi. *Multiple Access Protocols*. Springer, New York, 1990.

G. Romano, P. Salvo Rossi, and F. Palmieri. Optimal correlating transform for erasure channels. *IEEE Signal Processing Letters*, 12(10):677–680, 2005.

D. Sant. Throughput of unslotted ALOHA channels with arbitrary packet interarrival time distributions. *IEEE Transactions on Communications*, 28(8):1422–1425, 1980.

L. Schenato. To zero or to hold control inputs with lossy links? *IEEE Transactions on Automatic Control*, 54(5):1093–1099, 2009.

L. Schenato, B. Sinopoli, M. Franceschetti, K. Poolla, and S. S. Sastry. Foundations of control and estimation over lossy networks. *Proceedings of the IEEE*, 95(1):163–187, 2007.

L. Shi, M. Epstein, and R. M. Murray. Kalman filtering over a packet-dropping network: A probabilistic perspective. *IEEE Transactions on Automatic Control*, 55 (3):594–604, 2010.

B. Sinopoli, L. Schenato, M. Franceschetti, K. Poolla, M. I. Jordan, and S. S. Sastry. Kalman filtering with intermittent observations. *IEEE Transactions on Automatic Control*, 49(9):1453–1464, 2004.

J. Sommer and R. Blind. Optimized resource dimensioning in an embedded CAN-CAN gateway. In *Proceedings of the International Symposium on Industrial Embedded Systems*, pages 55–62, 2007.

P. Tabuada. Event-triggered real-time scheduling of stabilizing control tasks. *IEEE Transactions on Automatic Control*, 52(9):1680–1685, 2007.

A. S. Tanenbaum. *Computer Networks*. Pearson Education, New Jersey, 2003.

S. Uhlich and B. Yang. A generalized optimal correlating transform for multiple description coding and its theoretical analysis. In *Proceedings fo the IEEE Conference on Acoustics, Speech, and Signal Processing (ICASSP)*, pages 2969–2972, 2008.

G. C. Walsh, H. Ye, and L. G. Bushnell. Stability analysis of networked control systems. In *Proceedings of the American Control Conference (ACC)*, pages 2876–2880, San Diego, CA, USA, 1999.

X. Wang and M. D. Lemmon. Event-triggered broadcasting across distributed networked control systems. In *Proceedings of the American Control Conference (ACC)*, pages 3139–3144, Seattle, WA, USA, 2008.

X. Wang and M. D. Lemmon. Event-triggering in distributed networked control systems. *IEEE Transactions on Automatic Control*, 56(3):586–601, 2011.

Y. Xu and J. P. Hespanha. Estimation under uncontrolled and controlled communications in networked control systems. In *Proceedings of the 44th IEEE Conference on Decision and Control (CDC) and European Control Conference (ECC)*, pages 842–847, Seville, Spain, 2005.

W. Zhang. Stabilization of networked control systems over a sharing link using ALOHA. In *Proceedings of the 42nd IEEE Conference on Decision and Control (CDC)*, pages 204–209, Maui, HI, USA, 2003.

H. Zimmermann. OSI reference model – The ISO model of architecture for open systems interconnection. *IEEE Transactions on Communications*, 28(4):425–432, 1980.